APPEARANCE AND REALITY

APPEARANCE AND REALITY

AN INTRODUCTION TO THE PHILOSOPHY OF PHYSICS

Peter Kosso

New York   Oxford
OXFORD UNIVERSITY PRESS
1998

Oxford University Press

Oxford   New York
Athens   Auckland   Bangkok   Bogota   Bombay   Buenos Aires
Calcutta   Cape Town   Dar es Salaam   Delhi   Florence   Hong Kong
Istanbul   Karachi   Kuala Lumpur   Madras   Madrid   Melbourne
Mexico City   Nairobi   Paris   Singapore   Taipei   Tokyo   Toronto

and associated companies in
Berlin   Ibadan

Published by Oxford University Press, Inc.
198 Madison Avenue, New York, New York 10016

Oxford is a registered trademark of Oxford University Press

Library of Congress Cataloging-in-Publication Data
Kosso, Peter.
Appearance and reality : an introduction to the philosophy of
physics / Peter Kosso.
p.   cm.
Includes bibliographical references and index.
ISBN 0-19-511514-7 (cloth).—ISBN 0-19-511515-5
(pbk.)
1. Physics—Philosophy.   2. Relativity (Physics).   3. Quantum theory.
4. Realism.   I. Title.
QC6.K62   1997
530'.01—dc20   96-44182
                                                               CIP

Printing (last digit): 9 8 7 6 5 4 3 2 1
Printed in the United States of America
on acid-free paper.

CONTENTS

PREFACE

If you want to talk about the philosophical implications of modern physics, you really should know some physics first. That is the guiding principle for this book. There is a natural thrill in discussing the nature of physical reality, things like indeterminism, objectivity or subjectivity, inherent uncertainties, and so on. And so much the better if you can work a little quantum mechanics or relativity into the conversation. Sometimes though, the temptation to claim scientific authority gets ahead of the obligation to present scientific accuracy. That will not happen here. We will talk about all those philosophical issues, and we will see what modern physics has to say about them, but our conclusions will be based on a calm, level-headed understanding of the physics.

My goal is three-fold. First, I will present the foundational concepts of modern physics, that is, relativity and quantum mechanics. This presentation will focus on the logical relations of the core ideas of each theory and the pivotal experiments used to motivate them. Second, I will draw the philosophical implications from these theories, in particular the question of our ability to know about nature as it really is rather than merely as it appears to us. Finally, I will use all this as evidence to argue for my own ideas about reality and appearance, a position I call realistic realism. Modern physics, I will argue, gives us reason to believe that we can know some things about the objective, real natural world, but that does not mean we can know everything. You can see what is realistic about my realism.

You do not have to know any physics to read this book, but you must be willing to learn some physics. You don't have to know any mathematics to read this book, and you do not even have to be willing to learn any. We will get by without it. The presentation here is meant to be self-contained. It is not simply a survey of various people's ideas about philosophy of physics, or of various possible responses to philosophical questions. In the end I hope to prove my own point about realism, and the book is one long argument toward that end.

The argument is accessible to anyone interested in the intersection of philosophy and physics. This includes students of philosophy and students of physics. It also includes physicists with an urge to reflect on the more abstract ramifications of what they measure or calculate. And to anyone who has heard the sometimes alarming metaphysical claims allegedly sanctioned by modern physics, where all is uncertain or there is no world until we look, my argument should put things into perspective and calm the more outrageous misrepresentations of quantum mechanics and relativity. That is my hope.

I have had a lot of help in preparing this work in the philosophy of physics, and I appreciate it. Denny Lee has been my mentor when it comes to explaining physics in a way that anyone can understand. He showed me how much fun it can be, both for the teacher and the student. I hope that some of his inspiration and his style show up in what I have done here.

I have learned a lot from my students. They have helped me through some of the trickier concepts and forced me to clarify some murkiness that I would otherwise have lived with. Mitch Harris is the stand-out in this group. We have him to thank for Mitch's paradox.

Several people have read earlier versions of this book, and all have given me insightful comments that have made what you read better than what I first wrote. I am grateful to the reviewers for Oxford University Press, and to Paul Teller and Linda Wessels Winnie.

I thank Robert Miller at Oxford University Press for his perseverance and encouragement.

And thanks to Cindy. Every task is more enjoyable when she's around.

APPEARANCE AND REALITY

INTRODUCTION

⸱

> It is wrong to think that the task of physics is to find out how
> nature is. Physics concerns what we can say about nature.
>
> Niels Bohr

Bohr's assessment of the task and the abilities of physics can be put bluntly
in terms of appearance and reality. It is wrong, he claims, to think that
physics allows us to know the reality of nature. Physics can tell us only
how nature appears.

The distinction is important, since appearance, the human account of
things, is unavoidably influenced by our own perspective and preconceptions. The sun and moon, for example, appear to be roughly the same
size, but in reality the sun is much bigger. This much at least, we know.
In general, the way nature appears may be somewhat altered from the
way it in fact is. Appearance is affected both physically and conceptually
by the way we interact with things. The appearance of nature is therefore
incorrigibly subjective, but at the fundamental level of physics it is, according to Bohr, all we can really know.

It is not Bohr's opinion for its own sake that is of interest here. We want
to know the truth of the matter. What is it about physics that motivates
such cynicism, and, more importantly, is it warranted? Bohr was indis-
putably a giant in early twentieth century physics, but we should not believe his conclusion simply on the basis of authority when we can figure
it out for ourselves. That is what we are up to in this book, figuring out
whether physics can deliver knowledge of reality or only appearance. The
question is not of the proper task of physics in the sense of the most de-
sirable goals or the most aesthetic or useful results. It is not about what
anyone thinks physics *should* do. It is about what physics in fact *can* do.
What are the limits of knowledge in physics, constrained as it is to giving responsible proof for the claims it makes? How far can we responsibly go before we have lapsed into mysticism and the occult? If all of the
evidence of nature is essentially influenced by those who gather and apply it to knowledge, then all we can know seems bound to reflect this influence. We can speculate about the uninfluenced, independent reality,
but speculation without proof is not knowledge, and it is not physics.

This is the sort of reasoning that seems to leave reality, the way nature
is, beyond the limits of knowledge and outside the domain of physics. If
the argument is right, then as a matter of principle, the best that physics
can know is how nature appears to us. Bohr's case for this is based on
details of physics, both its scientific results and its methods. It is exactly
those details that we must look to in order to understand the distinction

between reality and appearance, and to see how far physics can go toward knowing how nature is.

It is only fair to point out that I have lifted Bohr's words out of context. He is in fact only reported to have said this about the task of physics. He never wrote it down, and so it may not accurately reflect his considered thoughts. I have taken one casual comment from a lifetime of work and reflection on nature and physics. This may be unfair to Bohr, but again, this not about Bohr. It is about the concepts of appearance and reality and the limits of knowledge. More precisely, the goal is to reveal how these concepts are clarified and related in twentieth-century physics. I admit to just using Bohr as a provocative and authoritative way to set things up and get us in the right mood.

Our goal is not to engage in the history of ideas but in the ideas themselves. Ignoring the historical context of Bohr's claim is not inappropriate, since what we want to do is put things into our own context, the context of today's physics. Given what we know about physics, the principles, the arguments, and the experiments, we shall see what is revealed about the issue of appearance and reality and about our ability to know either.

Bohr was a pivotal figure in the development of quantum mechanics. His philosophy took form during fast times in physics, under the influence of a new and strikingly counterintuitive description of nature. The same can be said of Albert Einstein, whose reflections on the distinction between appearance and reality come to a very different conclusion than Bohr's.

> Physics is an attempt conceptually to grasp reality as it is thought independently of its being observed. In this sense one speaks of physical reality.
>
> Albert Einstein

With Einstein, as with Bohr, we will not dwell on the personal motivation or the context of the ideas. Our interest is in the ideas themselves and the contrast they present for Bohr's account of physics. It is interesting to note though that these are the thoughts of the older Einstein, commenting on the accomplishments and disappointments of the new quantum physics. Einstein's theories of relativity were already safely in place. It is over the status and ramifications of quantum mechanics that Bohr and Einstein disagree. Einstein's vision for physics is more optimistic, holding out for knowledge of the way things really are rather than the mere appearances. And Einstein is no mystic. He does not intend for physics to speculate irresponsibly about physical reality and call that scientific knowledge. Einstein's claim is that we can know in a responsible way, with proof, about nature itself and not just our subjective description of nature. Physics offers objective knowledge in this sense, knowledge that transcends the appearances.

This philosophical question of appearance and reality warrants a responsible answer, one supported by proof rather than faith or blind hope or authority, no less than does any scientific question. This will be the nature of our inquiry. But what is the source of evidence for such a philosophical question, and what would be the nature of the proof? We are looking for a general account of the limits of knowledge in physics. Will we find what we are looking for in physics itself, that is, by doing what physicists do? Is there some experiment or some set of calculations that will clear this up? Or are we faced with a purely philosophical question, one that can be framed and addressed only by stepping outside the enterprise of physics and looking back?

For Bohr and many others who have come this way since, the answer lies within physics. The physics of quantum mechanics forces the hand of philosophy here, just as the evidence from the lab constrains the scientists' theorizing. The principles and experiments of quantum mechanics themselves show the truth about appearance and reality and our ability to know. The quantum world is one of essential uncertainties and irresolvable probabilities. It is described in terms of complementary, that is, contradictory, properties. Things exist somehow as waves and as particles, and yet as neither, an ambiguity that is resolved only by our intervention in the act of observation. Definitive properties such as being a wave or being a particle, even being in any particular place, seem to be created by the observer. The things we know are apparently as much our own doing as the world's and we seem to be stuck describing how nature appears to us rather than how nature is in itself.

In this way the details of physics function as evidence for the more general and abstract philosophical claim about knowledge. This is the strategy we will use to resolve the main issue of appearance and reality. What does physics itself indicate about the nature and relation of appearance and reality? And most importantly, what can we know about each? In terms of our ability to know, what is our relation to the physical reality of which Einstein speaks? Does physics itself favor either Bohr's or Einstein's account of what we can and cannot know?

The philosophy is in the details of the physics, and so the plan must be to first understand the foundational concepts of relativity and quantum mechanics. These are the conceptual heart of contemporary physics. Relativity is the basic framework for describing the nature of space, time, and motion. Quantum mechanics is the received account of the basic constituents of the stuff of nature, and of interactions that exchange energy and information from one small part of nature to another. Relativity and quantum mechanics represent the essence of how physics is done and the manner by which physics describes nature. There is nothing for it then but to come to terms with the basics of these aspects of physics.

To say we will study the basics of relativity and quantum mechanics does not mean we will skim these theories for the easy parts and settle for a shallow cartoon of the issues. We will get to the conceptual foundations, as we must for a responsible solution to the philosophical question, and we will get there without mathematics. In place of the often perplexing squiggles of algebra and calculus, much of the conceptual core of relativity and quantum mechanics will be shown with the help of some simple diagrams. We will also avoid the inscrutable jargon that can block access to the story physics has to tell. The story will not always be easy, but it can be told in plain English. On the few occasions where efficiency is served by a technical term, a clear explanation will come first.

Once the basic concepts of relativity and quantum mechanics are in place, we will clarify the philosophical ramifications for the questions about appearance, reality, and knowledge. It is essential to get the physics right before asking any philosophical questions. It will not do to create a caricature of the science with some features unnaturally exaggerated and others hidden to allow a pet interpretation to emerge. This would be no less dishonest than fudging the data in an experiment. Accurate and detailed physics is the prerequisite as evidence for any valuable philosophizing on this issue.

Any effort to gather evidence is more valuable with some preliminary planning and conceptual groundwork to know what we are looking for, what it will look like, and how to find it if it is there to be found. For this reason, the first two chapters will provide some general philosophical background in preparation for the physics and its meaningful analysis.

Chapter 1 discusses the relation between physics and philosophy, and asks which is the proper forum for addressing the issues raised by Bohr and Einstein. More generally, it surveys the kinds of questions and issues normally found within each discipline, and describes some differences and similarities in the standards of argument and proof in each. The discussion in the first chapter leads to several foundational issues that are shared between physics and philosophy. It closes by clarifying some helpful philosophical distinctions that can be used to expedite the project of understanding the nature of appearance and reality in physics.

Chapter 2 is more specifically about the concepts of appearance and reality and the question of knowing nature. The primary goal is to pose the question in a clear way that is amenable to an answer. We will clarify what is meant by the key terms such as "appearance," "reality," "objective," "subjective," and the like. This is the stereotypically philosophical advice to define your terms before attempting an answer. In this case it is good advice because we are dealing with concepts that are so deeply intuitive that many of us have lost the ability to articulate their meaning. By noting the variety of senses of these ideas, and settling on what is most appropriate to our philosophical concerns, we will be doing the conceptual work of clarifying just what to look for in the evidence. We are, in this sense, formulating part of the hypothesis necessary for a meaningful experiment.

Then we will do the physics, with three chapters on relativity and two on quantum mechanics.

Chapter 3 presents the essential, foundational concepts of the special theory of relativity. The medium of description will be plain English and Minkowski spacetime diagrams. These diagrams are straightforward and handy ways to represent the spatial and temporal properties of objects in motion. They show at a glance the effects of different perspectives from different frames of reference, and this is surely relevant to the issue of appearance and reality. The Minkowski diagrams will do the work that is often done in the algebra of so-called Lorentz transformations. We can do all that is important, and do it in a clear and efficient way, with pictures rather than equations. Chapter 3 will not rest until we have a solid understanding of the foundations of the special theory of relativity. Starting from the defining principle of relativity, that the laws of physics are the same in all non-accelerating reference frames, we will survey the logic and evidence that lead to the perplexing claims that the speed of light is always the same, regardless of the speed of the source, and that nothing can go faster than light. These are the basic principles of the special theory of relativity from which follow surprising accounts of the nature of length of ordinary objects, of time duration and simultaneity. All of these pieces of the special theory of relativity will be put into place in Chapter 3.

Chapter 4 takes up the general theory of relativity. The structure of exposition will be like that of Chapter 3, first presenting the basic principles and then using them to give a coherent account of the nature of space, time, and, in this case, gravity. This is the *general* theory of relativity because the foundational principle insists that the laws of physics are the same not just in all reference frames that move at constant velocity, as is the restriction in the special theory, but in all reference frames, period. Even on rotating carnival rides and accelerating airplanes where it seems that invisible forces toss things around, the laws of physics take the same form as in a peaceful room at rest here on the earth. In the peaceful room, after all, there is this invisible force of gravity that draws things to the floor, not unlike the apparent, invisible force that draws things to the back of the airplane as it accelerates down the runway. Following this line of reasoning leads to an account of gravity in terms of motion and the properties of space and time. We will talk about the curvature of spacetime with careful attention to what is measurable. Under the influence of the philosophical question about the limits of knowledge, we will ask whether it is more accurate to the theory to say that spacetime *is* curved or that spacetime *appears* curved.

Chapter 5 returns to the philosophical issue and collects the consequences of the exposition on relativity. It will be important to do this before encountering the more divisive quantum mechanics. The experience gained in interpreting relativity might help us resist the amazing but unwarranted interpretations that often come in the wake of quantum mechanics. In Chapter 5 we will want to clarify the sense in which relativity tells us how nature is and in what sense it is only about how nature appears.

Chapter 6 gets back to physics by describing quantum mechanics. The quantum model of nature and the basis of its plausibility are shown both in the foundational principles and the important experiments that advanced both the discovery and justification. Principles of complementarity and uncertainty will be explained, as will the nature of measurement of a quantum system. A great deal of philosophical mileage will be gained from the famous Einstein-Podolsky-Rosen thought-experiment and its extension by Bell's theorem. All of these aspects of quantum mechanics will be described in sufficient detail to give a genuine appreciation for the distinction between the quantum and classical descriptions of nature.

Chapter 7 then draws the philosophical conclusions about quantum mechanics, just as Chapter 5 did about relativity. Here we will encounter the Copenhagen interpretation, named for the city where Bohr worked. We will also discuss less mainstream accounts of the quantum world such as those that give human consciousness an active role in determining nature, and the many-worlds model. This is to be more than just a survey of alternative interpretations. The idea is to see just what philosophical position is warranted by the physics of quantum mechanics.

The final chapter summarizes and generalizes the results of the previous seven and puts things together to answer the question of knowing appearance or reality. The philosophical results are summarized under what I will call realistic realism. It is realistic in the sense of being practical and explicitly admitting of our limitations. We can know some things about the way nature is, but not everything. The details and the proof are what the whole book is about.

Chapter 1
PHYSICS AND PHILOSOPHY

Physics and philosophy sections are usually at opposite ends of the bookstore. While physics presides over the hard sciences, philosophy is somewhere beyond even the softer sciences, past psychology and to the left, against the wall with those tenuous, speculative topics like the occult and spiritualism. This is an unfortunate floor-plan because it gives a misleading impression of the relation between physics and philosophy. They in fact have a lot in common. The comments by Bohr and Einstein, and much of the science we will be doing in what follows, make it clear that physics and philosophy intersect on some significant issues. Once we get our hands dirty with discussions of space and time, cause and effect, determinism, and the like, it will be hard to tell whether the dirt came from physics or philosophy. It will also be hard to tell whether the methods we are using—whether it is to get dirty or to get cleaned up—properly belong to physics or philosophy. There is common ground not only in what questions the two disciplines ask but in what standards they require of acceptable answers. Plausible philosophy, no less than plausible physics, is based on evidence and reason.

There are, of course, differences between physics and philosophy, but there are also similarities. We need to clarify both, because the issue of appearance and reality is shared between the two disciplines.

WHY THIS WILL REQUIRE BOTH
PHYSICS AND PHILOSOPHY

No one appreciates an outsider meddling, and perhaps getting all philosophical about physics would be an annoying disciplinary indiscretion. Before rearranging the bookstore, we should ask whether one discipline or the other can handle the question of appearance and reality on its own. It is best not to frame this question in terms of the people in either discipline or their departmental affiliations. Plenty of people who have been trained as physicists, who call themselves physicists, and who work in physics departments address philosophical issues in philosophical ways. The same is true of philosophers acting like physicists. What is important is not who does the work or in what disciplinary setting, but how the work is done. The relation between physics and philosophy that is important here is in terms of the methods and standards of argument used by each. We need to talk about comparisons between physics and philosophy in terms of what counts as evidence and proof of a claim. This will help to clarify what kind of argument, what kind of proof, is appropriate to the question of appearance and reality.

Perhaps Bohr has raised an issue that is strictly an internal matter for physics. This would demand that it be settled, as best we can, by scientific methods and scientific standards. This is not to say that the correct resolution is simply what the scientists themselves say about it. It is not an opinion poll. There is more to science than the people and their beliefs; there are standards of evidence that are the basis of authority of any claim. Can we apply the same standards to the question of appearance and reality and the capacity of physics as we do to finding the cause of AIDS or the age of the earth?

Perhaps, on the other hand, Bohr's claim about physics and knowledge requires an external judgment. At issue, after all, are the limits of scientific knowledge, and we may invite circularity into the argument if we use scientific knowledge to judge the abilities of scientific knowledge. An internal argument about physics will be well informed, but it might also be self-serving. It is not that the people who do physics cannot be trusted to tell the truth on the limitations of their work. No, it is not like asking the Defense Department to evaluate its own weaponry. The concern here is of a more deeply logical circularity of using the methods of science to evaluate those methods. The problems this entails could be so subtle and implicit as to go unnoticed by even the most forthright and well-intentioned scientists.

Because of this concern over leaving the issue of appearance and reality as strictly internal to physics, we are forced to say that Bohr's claim takes us beyond physics. It is an issue *about* physics. But this does not mean that it is strictly an issue of philosophy. Without some honest evidence from the experiments and principles of physics, the resulting philosophy would belong back there against the wall with the astrological guidebooks. The question of appearance and reality *necessarily* involves both physics and philosophy.

It is not simply a matter of convenience or efficiency or diplomacy that we look to both physics and philosophy on the issue of appearance and reality; it is a necessity. Just as any scientific account of nature requires information from both theory and experiment, any meaningful and credible account of the ability of physics to know appearance or reality requires information from both physics and philosophy. In science, experiment without theory is meaningless, since one needs to know both what to look for and, once it is found, what it indicates about interesting but unobservable things like atoms or germs. Furthermore, theory without experiment is idle speculation, since experiment compares the grand generalizations of theory to the particulars of experience. It is the same on the issue of appearance and reality. Philosophizing with no regard for the details of physics would be the idle speculation, while physics with no philosophical framework would be, for these purposes, meaningless. Adapting a line from Kant, admittedly a philosopher, philosophy without physics is empty; physics without philosophy is blind.

Linking physics to philosophy does not mean that the issue that Bohr raises and we hope to settle is one of opinion. It has nothing to do with what anybody would like to see physics aim for. It is not even about what anybody believes physics can accomplish. Asking just what someone believes on the issue is not the business of philosophy. The bottom line here is what is *in fact* knowable in physics. What, in fact, can be proven given the methods of physics and the limitations of human beings? The "proven" here should not mean proven with dead certainty, for then no scientific claim would qualify. It is more realistic and more profitable to pursue some degree of responsible justification that separates knowledge from guesswork and dogma.

Asking about the limits of scientific knowledge is a lot like asking about the resolving power of a microscope. Perhaps there is not an exact demarcation between sizes of things that can be resolved and sizes that cannot. The distinction may not be sharp, but it is not a matter of opinion. There is a fact of the matter as to the capabilities of a microscope, as there is about the capabilities of scientific knowledge in general. Furthermore, the power of a microscope cannot be determined just by using it. We can look into the eyepiece and note which images are clear and distinctly resolved, but this alone cannot tell us the size of the specimen. We can, in this sense, know what we are seeing but not what we are looking at. To know how good the microscope is we will need to know some more general principles regarding how it works. To know the limits of knowledge in physics we will need to know some more general things regarding how it works.

STANDARDS OF PROOF

The difference between opinion and proof, between speculation and knowledge, is a concern that is shared by physics and philosophy. Both pay explicit attention to the standards by which to tell the difference between speculation and knowledge, the standards of proof, and it is this attention that brings physics and philosophy together on Bohr's question.

A central part of the scientific method, the method of physics, is a set of standards for monitoring and controlling the circumstances of gathering information about nature. In the lab, experiments and observations must be done with care, with the appropriate controls and under the proper conditions. The experts, the scientists, have an idea what would distort the information, and so they know what needs to be controlled and what counts as proper conditions.

The expertise for gathering information is applied at various levels of generality in physics. In a particular lab, running a particular experiment, the participants must come to know the vagaries of their particular apparatus and specimen. They know just how to warm up the machine so that it works properly, and they know where to tap it when the needle sticks. On a somewhat wider scale, in a specific area of physics such as surface science or super-conductivity, the group of participants shares the background knowledge relevant to credible observation. This is the basis of peer review of proposals and results, a common knowledge of the applicable theories and standards for running experiments and gathering information. Everyone in surface science, for example, knows about baking the experimental chamber to make sure the sample is in a good vacuum.

At an even broader level of generality, all physicists, regardless of their special interests, share basic theoretical and methodological beliefs that guide the acquisition and interpretation of data. An understanding of basic conservation laws and the nature of interactions is an essential component of the standards for judging any observation in science. Thus, any physicist has the necessary expertise to evaluate at a basic, general level, any informational claim in physics. And this kind of evaluation is essential to the discipline. Keeping careful track of the conditions of gathering information is part of what makes science scientific.

Philosophy is in a similar business of evaluating the plausibility of information and assessing the responsibility of proof. Philosophy though is conducted at a level that is one step more general than physics as a whole. Philosophy reflects on principles of scientific method in general, still with an eye on the circumstances of gathering information. At the philosophical level of generalization, the focus is on the interaction between scientific theory and evidence and how each influences the other. Generalizations about this kind of interaction must be based on specifics from particular cases of observation and theorizing in science. Just as any generalizations astronomers make about pulsars must be based on observations of a few pulsars, philosophical claims about physics and the nature of scientific knowledge must be based on a detailed account of physics in action. Philosophy relies on this kind of evidence.

Questions of accountability of information and justification of knowledge thus shade from physics to philosophy. And of all the sciences, physics comes the closest to the level of generality in philosophy. Physics aspires to make claims of unlimited universality. Unlike biology, which is only about things that are alive, and unlike anthropology or psychology, which are even more narrowly only about people, physics is about everything. Alive or not, human or not, all things are subject to the laws of gravitation. All things are made of atoms. Physics describes the most basic and ubiquitous elements of nature. It deals with the fundamental interactions that are the synapses of nature, all of nature. This universality gives physics a level of generality that brings it close to that of philosophy.

FROM PHYSICS TO PHILOSOPHY

There is another way, besides sharing general concerns about the accountability of knowledge, that physics leads to philosophy. Many of the fundamental claims of physics are profoundly counter-intuitive. This is nothing new with the twentieth-century revolutions. Physics has always been this way, regularly defying both common sense and basic observation.

Galileo made his mark by claiming that, contrary to the way things feel and look, the earth is moving while most of the objects in the heavens are standing still. The earth both orbits the sun and spins on its own axis at a pretty good clip. It may look as if the sun moves up at dawn, but the new physics revises this observation. Galileo also claimed that an object in motion will continue at the same speed and in the same direction if there are no outside forces acting on it. In other words, it does not take any force to keep something moving. This account of motion, made precise and mathematical by Newton, is a fundamental truth of physics. It is also contrary to the way things seem to work. Turn off the force from the engine and your car slows to a stop. It *seems* as if the force is required to keep the car moving. A sled will slide when you pull it, but on its own it comes to a stop. Again, Galileo's idea of motion without force seems wrong. Of course the problem in both cases is that we have not accounted for all of the forces on the car or sled. Friction is always at work, and that is what stops us. If there were no friction, and hence truly no force on the car or sled, they would continue to move forever. But all of this has to be told to you. It is not common sense.

The atomic theory is similarly defiant of basic observation and common sense. Solid objects like a table or a stone may look and feel both continuous and densely packed with material, but in fact they are grainy assemblages that are by far mostly empty space.

Given that many basic claims in physics fly in the face of both common sense and our natural observations of events, the question of the status of such claims arises. Are these principles of physics dead certain? Are they only likely, to some accountable degree, to be true? Or are they optional parts of a descriptive model that affords convenience and practicality in dealing with nature but that has nothing to do with truth or accuracy? The more useful and more important way to ask this kind of question about status is the more general, more philosophical. How do claims in physics achieve their status of being certainly true or probably true or whatever? What are the standards used to achieve such a status?

I should not pretend that I am going to answer these questions directly. The point in raising them now is to show how basic physics leads directly to basic philosophy, and that an understanding of the physics is enhanced by an understanding of the philosophy.

Since the examples we have seen of foundational principles of physics actually defy our day-to-day, unschooled observations of nature, we must doubt that observation by itself can be the basis of proof for scientific claims. The role of observation in the scientific method must be more complicated than just having a look and seeing what is happening in nature. Unschooled observation actually tells against these scientific claims. Trained and interpreted observation, on the other hand, must count somehow as evidence in support of the challenging science. New theories give us new ways of looking at nature. But how exactly does theory reinterpret observation? That is, if unschooled observation does not serve to support the claims of physics, what sort of schooling does it take to be of service, and how is this schooling related to scientific theory, perhaps the very theory that the schooled observation is supposed to prove? These are philosophical questions, brought on by the style of the most basic claims in physics. The counter-intuitive and observation-defiant nature of physics calls for a general description and evaluation of the scientific standards of evidence.

If pure observation is not the source of verification for scientific claims, then perhaps pure reason is. A scientific claim must make sense, and when stitched together into a tapestry of ideas that is a whole theory, the network must be coherently reasonable. But physics often does not make sense or seem reasonable to those whose sensibilities have not been prepared by years of education. Again, unschooled reason does not always believe the outlandish claims of physics, and we must wonder about the schooling and its effect on the objectivity of reason. Sometimes physics does not make sense even to the experts. Richard Feynman, a key contributor to the development of modern physics, put quantum physics in this light: "I think it is safe to say that no one understands quantum mechanics." We will see for ourselves how profoundly challenging quantum mechanics is to our common sense. The world of quantum physics will seem to put us in the perplexing situation of the Omniscienter in Rene Daumal's *A Night of Serious Drinking*. The sign over his chair reads, "I know everything, but I don't understand any of it."

The point, if Feynman is right, is that making sense cannot be a basic standard of credibility in science, and the faculty of reason alone cannot be the source of scientific justification.

All these questions about the relation between observation and theory, and the possibility of foundations of scientific knowledge, are at a level of abstraction and generality to make them philosophical questions. They were brought on by the challenging character of the principles of physics, and in this way, physics leads to philosophy.

USEFUL PHILOSOPHICAL CONCEPTS

Since we are among philosophical questions, efficiency will be served by agreeing to use just a few specialized philosophical terms.

I will make free use of the term "epistemology" to mean the study of knowledge. As biology is a domain of inquiry about the nature of living things, epistemology is a domain of inquiry about the nature of knowledge. We have been asking epistemological questions already without really trying to answer them. The question of proof is an epistemological question. The issue of observation and evidence and their relation to theory is epistemological. And so is the main question about the limits of knowledge. Can what we observe function as evidence to justify claims about what cannot be observed? Can knowledge, in other words, transcend observation? This is an epistemological question. It gets directly to the details of using the appearance of nature as evidence for knowledge of how nature really is.

Adopting a special term for this kind of questioning is useful for focusing attention on the kind of answer that is expected. It helps in avoiding misunderstanding and arguments at cross-purposes. If the discussion is epistemological, the issue is *how* we know about nature; it is not about *what* we know about nature.

The important contrast to epistemology is the domain of inquiry about the general nature of things. This asks what nature is like, without regard for how we know about it. There is a choice of terms for this kind of study. Some people call it ontology, others call it metaphysics. I opt for metaphysics as a general term for issues of existence. Thus, metaphysical questions are about the way things are. The label metaphysics may invite confusion, as in the metaphysical/occult section of bookstores. But we are already rearranging the bookstore to suit our interests. As the term is used here, it describes a kind of question, not a method of inquiry or standards of proof or a kind of answer. It means only that we are asking what there is in nature and what it is like. Metaphysics in this sense is part of the common ground between physics and philosophy.

We have already come close to some metaphysical questions without explicitly asking them. Is there a determinate way that nature is, independent of our looking at it or thinking about it? Or do we create the natural world as we interact with it? These are metaphysical questions. So are the less general but equally abstract questions that will arise about particular aspects of nature such as space and time. Are space and time continuous or do they exist like sand with discrete grains of place and moment? We can ask whether the events of nature are strictly determined so that from one moment to the next there is no element of chance or probability. This question of determinism and probability has an epistemological cast as well. Events could be deterministic in fact, but, because of the limitations of our ability to observe and understand, we cannot know what will happen next but can only assign some degree of likelihood. In this case, probability is a purely epistemic occurrence, not metaphysical.

The terminology and explicit contrast between epistemology and metaphysics are introduced to avoid confusion in addressing Bohr's issue. Bohr uses quantum mechanics to argue that the way the world appears is relative to the observer. Perspective makes an indelible contribution to the description of nature. This is an epistemological claim, because it describes an aspect of our knowledge of nature. What we know depends on us. But it does not necessarily follow that the way things are depends on us. This is a metaphysical claim and it is separable from the former question about the nature of knowledge. The human influence on the knowledge of things does not necessarily imply a human influence on the nature of things.

It will always be helpful to separate the epistemology from the metaphysics. We want to avoid answering epistemological questions with metaphysical claims.

PHILOSOPHICAL ISSUES

The distinction between epistemology and metaphysics helps to clarify some of the philosophical issues that will show up soon. The issue of realism, for example, can and should be separated into epistemological and metaphysical forms both in the questions asked and the answers given.

The question of metaphysical realism is of existence: Is there a reality that is independent of us? One possible response to this, the one that says yes, there is, is itself given the label metaphysical realism. The opposite position, that there is no reality independent of us, that the way things are is dependent on our actions or observations or conceptions, gets no more distinctive a name than metaphysical anti-realism. Quantum mechanics has made metaphysical anti-realists of many physicists. David Mermin, for example, expands on the quantum results to say, "The moon is demonstrably not there when no one looks." This and similar metaphysically anti-realist claims warrant clarification and questioning. That will begin in Chapter two. The task now is simply to separate questions of existence from questions of knowledge.

There is also an epistemological realism, both as an issue and as a particular philosophical position. The issue is whether we can know an independent reality and not just appearances. Can we infer from the way things appear to the way they really are, from claims that the world is *as if* this and that, to claims that it *is* this and that? The position of the epistemological realist is to answer yes, we can. This bears a burden of proof to show how claims to know about an objective reality can be justified by subjective, perspective-influenced appearances. Epistemological anti-realism would be the claim that this requirement of proof cannot be met. That is, a great divide separates the information we can access in appearance and the information we would like about reality, and there is no way across the divide. This anti-realism is an easier position to defend than the epistemological realism, because it requires no proof of its own. It is enough to point out the shortcomings in the epistemological realist's attempt to bridge the gap. But the epistemological anti-realist is in a curious position with respect to the metaphysical issue of realism. A responsible epistemological anti-realist cannot take either side on the metaphysical issue. If we cannot know anything about nature beyond our own observations, then we cannot responsibly make any metaphysical claim, either that there is or is not an objective world beyond our observation. A kind of metaphysical agnosticism seems the most appropriate complement to epistemological anti-realism.

The question of objectivity also has both a metaphysical and an epistemological form, and they are worth separating. The metaphysical issue is discussed in terms of an objective world or objective reality. The question is of the nature of things and its independence from any particular person or people in general. An objective property of a thing is one that is not a matter of opinion, or taste, or perspective, or creative invention. Contrast the non-objective claim that this water is cold, with the objective claim that this water is 60 degrees Fahrenheit. To the former, anyone can respond, "No it's not!" and there is no way to say that one of us is right and the other is wrong. There is no fact of the matter beyond the appearance. It is just a matter of personal perception, how it seems to the individual, and being cold is an entirely subjective property. But there is a fact of the matter regarding the Fahrenheit temperature of the water. This is an objective property.

The epistemological issue of objectivity has to do with standards of testing and proving a claim about the world. It is what we have in mind when we talk about an objective jury. It is not that they exist independently of us that makes them objective, but that they decide what to believe in a way that avoids personal bias. Their objectivity is a feature of how they come to a verdict, how they come to know the defendant's guilt or innocence. It is an epistemological sense of objectivity. Extending this sense of objectivity to science requires that the methods of scientific inquiry avoid the bias of a whole disciplinary group and their theoretical or methodological predilections. The value in this kind of objectivity is that, by escaping our own point of view, we are more likely to arrive at the truth of the matter.

Once again, the reason for distinguishing the epistemological from the metaphysical questions of objectivity is because these issues will come up later, in the heat of the physics, and it will be helpful to be able to clarify which version we are working on. In this way, a little philosophical analysis gives a framework for addressing questions about scientific knowledge.

PHILOSOPHICAL EVIDENCE

Scientific theories need observational evidence to back them up and stimulate revisions. Philosophical claims need evidence too, for exactly the same reasons and in exactly the same way. Epistemological generalizations, for example, must be based on the specifics of actual cases of knowledge to add plausibility and practicality to the general framework. Real cases force revisions if the philosophical model does not fit or does not help us explain and understand the specifics of knowledge.

The requirement of philosophical evidence is an aspect of what is sometimes called naturalized epistemology. The general attitude in this approach is to avoid a purely abstract philosophy based only on logic and reason, and that would prescribe how knowledge must be done by generating a model of successful knowledge to which real cases must conform. Naturalized epistemology is empirical, based on evidence of how knowledge is in fact done. Perceptual knowledge, for example, can only be understood by studying the details of optics, physiology, and psychology. The role of experiments in scientific knowledge, for another example, cannot be analyzed in any abstract way but only by seeing how real experiments work and by studying the theories and machines involved. Understanding knowledge, by this strategy, requires understanding the physical nature of the knower and the things known. In other words it requires knowing the metaphysics of the interaction between humans and the world.

Naturalized epistemology learns what we can know and how we can know it by first noting what the nature of things, both ourselves and the world, allows us to know. Science is most often the source of information on the nature of these things. In our case it will be what physics can tell us about the limits of knowledge. But this is again flirting with the circularity, using knowledge in physics to evaluate the limits of knowledge in physics.

There is a similar circularity that threatens within any science. Scientific evidence must be accountably credible and meaningful. Often the accountability relies on some theoretical account of how evidence is acquired and what it means. But then scientific evidence is used to test the accuracy of theories. Scientists must be careful that it is not the same theory that benefits from evidence as was used to support the evidence. The same sort of care must be applied to philosophical evidence in naturalized epistemology.

Since the evidence for any philosophical theorizing about scientific knowledge must be in cases of scientific knowledge, the relation between philosophy and physics closely resembles the relation between theory and evidence within science. In the scientific context and in questions about knowledge and realism, some preliminary conceptual framework is needed to direct the selection and interpretation of the evidence. Both theory and evidence are necessary in any scientific issue, and both philosophy and physics are necessary to make progress in the issue of realism.

So we must do some philosophy before the physics. In the next chapter, Bohr's distinction between appearance and reality will be clarified, and the question of what physics can know will be posed in a way that is amenable to resolution by the evidence.

Chapter 2
APPEARANCE AND REALITY

Standing waist-deep in clear water, your legs look shorter and fatter than they really are. A straight stick seems to bend where it enters the water. In these cases there is a difference between the way things appear and the way they really are. We can appreciate the difference here because we know that the stick is really straight and your legs are long and slender. We know the reality and can compare it to the appearance. Bohr's claim though is that in the context of foundational physics we cannot know the reality behind the appearances. We can know nature as it appears to us but not as it is.

Two aspects of this central theme need clarification. In the first place, why should we think that we are limited to knowledge of appearance and precluded from knowledge of reality in the case of foundational physics? With sticks in the water we got beyond the appearances; why can't we do that in all cases? Second, in those cases where we cannot transcend the appearances, why think the reality is any different from the way things seem? If all we know are bent sticks, why think there is any more to it than bent sticks?

SCIENTIFIC OBSERVATION

Understanding the difference between appearance and reality will be furthered by a discussion of the nature of observation. There is no deep philosophy or physics required to note the obvious effects of perspective on the results of observation. Because of their different distances, a hiking companion can appear to be as tall as a mountain. Your friend's head seems to fit easily between your own fingertip and thumb. But we know she is much bigger than that, just as we know the mountain is much bigger than she is.

One of the keys to recognizing a difference between appearance and reality in this case is that as you move about and change your perspective, the scenery and your friend grow and shrink and change size even with respect to each other. The observation is obviously influenced by the circumstances of the observer, and so what you see depends in part on you. What you see is not the way things are in themselves, but the way things are from your perspective. Perceptions are also episodic in the sense that your friend disappears when you look away, shows up again when you look back, and so on. But reality is not episodic in this way. People and mountains do not blink in and out of existence, just as sticks do not really bend when dipped in water.

Once the effects of perspective are understood, they can be taken into account and reality can be reconstructed out of the appearance. That is, knowing the circumstances of observation, we can figure out what the appearances mean about reality. This is the essence of scientific observation. A *scientific* observation is more than a casual glance. It is not just a wide-eyed, gullible, seeing-is-believing look at nature. It must be more knowledgeable observation that considers the circumstances. Thus, your friend looks to be as tall as a mountain, but with a little trigonometry you can figure out just how high the peak *really* is.

In the context of science, appearance must mean more than the report of a thoughtless, careless glance. It must be the most careful and considered look. It must be the contents of observation, all things considered, so that we are not fooled by the vagaries of perspective. But this consideration, intended to lead observation closer to reality, might have just the opposite effect. The addition of thoughtful consideration endangers the objectivity of the project, since it introduces more human influence rather than less. If we think about and interpret every observation, we may end up seeing only what we want to see.

This quick reflection on the prerequisites of scientific observation is enough to show that the observer influences what is observed. And since the influences are inescapable (the glasses are always tinted, so to speak), there is reason to doubt the realist's belief that we can know how nature really is, clear and untinted. Surely the realist bears the burden of proof. And certainly a naive realism of the sort that says the appearance is a direct and exact mirror of reality is untenable. Any credible claim to know reality will require a critical and careful reconstruction, using the appearance as evidence. The reconstruction will require taking into consideration the influence we have and the information we add to the appearances. The human influence on observation has two components, one physical and the other conceptual, and we must attend to both. In subsequent chapters we will direct the attention of physics to this question and see what the science itself says about the possibility of knowing how nature is.

THE CONCEPTUAL INFLUENCE

Scientific observation must be done in a way that is careful, meaningful, and accountable to possible objections. To function as evidence, an observation must be more than a private, physical act of sensation. It must be an epistemic act in the sense of being well founded and well understood enough to contribute to knowledge. It must be described in a language that is relevant to other things we know. All of these requirements, from being careful to being meaningful, depend on some antecedent background knowledge. What we already know influences what we observe.

There are three distinct ways that this conceptual influence happens in science, and they are all unavoidable.

The first has to do with selecting which things to observe and which to ignore. It is a matter of deciding where to shine the light on the vast darkness that is nature. Gathering data in science is neither comprehensive nor random. We cannot look at everything. Nor can we waste time by looking haphazardly at one damn thing after another. Scientific observation must be selective and methodical to be sure we see what is important and relevant to our interests. And any determination of relevance will be directed by the current theoretical understanding of nature. Theories, after all, tell us what is relevant to what by describing how nature is put together and what causes what.

The situation here is similar to conducting a political opinion poll of people, similar in that the survey of data must be representative. Selecting an uncharacteristic sample will produce a deceptive view. The evaluation of the data for proper representation is based on regulating the relevant factors. You want to question some men and some women, some young and some old, urban and rural, and so on. These features are relevant to political opinion. But you do not care if some of the people chew gum and some do not, or some are short and some tall. These factors are irrelevant. But this determination of relevance, insofar as you have to know it prior to observation, is under the influence of our conceptual sense of the situation. We must already know some general (theoretical) things about political opinions before we can gather meaningful, credible data on political opinions. What we already believe about things directs what to look at.

The second requirement of conceptual influence on scientific observation is in judging the credibility of observation. Sometimes our senses deceive us. Sometimes the instruments of observation, like microscopes or thermometers, malfunction and give inaccurate information. A credible observation must come with assurances that could be given, even if they are not explicit, that these sorts of mistakes are not happening now. We cannot leave it to chance that the conditions for viewing are right and the machines are working properly. We must know the right conditions from wrong, and here again is a role for background knowledge, this time regarding how the observation itself happens. Like judging the resolving power of a microscope, there is no way to tell just by looking that it is casting an accurate image of the specimen. We need to know how it works in order to know that it is working properly.

The fact that observations are not automatically credible shows up in the pragmatics of the scientific method. If an observation contradicts a theory, the first tinkering to be done is to check and make sure the observation has been properly performed. At least part of that checking will call on a conceptual understanding of how the observation ought to be performed.

The third role of theory in scientific observation is in knowing what a particular observation means. To function as evidence, an observation must be of something that is relevant to a theory. The streaks of vapor in a cloud chamber or a bubble chamber are of no interest at all unless they can be linked to the passage of elementary particles. A certain pattern of streaks might mean an alpha particle with a particular energy. The evidence must be of particles, not just streaks, and to make the connection some theoretical understanding of how particles produce streaks is required. To render the observation in a theoretical vocabulary will require some theoretical input. So here again, the elevation of observation from the brute physical event of sensation to the useful epistemic event of evidence will be done under the influence of background theory.

This is the Kantian theme again, that experience without concepts is blind. Our encounter with nature is meaningless without some organizational network of concepts into which we fit our sensations. And this makes it seem as if we are bound to know nature on our own terms and within our own conceptual arrangement rather than on any natural terms. The conceptual perspective is inescapable and appearance is always, as they say, conceptualized.

It would be irresponsible to end the discussion here, like convicting a defendant at the arraignment. It is still a fair question whether it is possible to transcend these appearances to know an unforced, unaltered, natural reality. But we want physics to answer that. Right now we are just clarifying the questions and casting doubts.

THE PHYSICAL INFLUENCE

In addition to the unavoidable role played by the observer's antecedent knowledge and conceptual habits, the observer will also have a physical influence on the thing observed. In addition to the effects of perspective, the physical act of observing changes the properties of the object itself. This will happen to different degrees in different cases, and it will be important in each case to clarify how the physical intervention changes the object and its appearance.

The effects of the observer on the observed are well known in the social sciences. It is a central methodological problem for anthropologists, since any group of people is likely to act differently when a stranger is among them, a stranger who is obviously keeping careful track of what they do and say. The anthropologists can do their best to keep a low profile, but they cannot disappear.

Psychologists face a similar challenge with individuals. The goal is to find out what someone is thinking or what is going on in the subconscious. The method is in asking questions and listening to the answers. The risk is in suggesting answers in the questions themselves. Even without the questions there is an artificial aspect to the analysis in that people are likely to behave and speak differently when they know that their words and actions are being analyzed. They may not be themselves, being perhaps more guarded than normal. The doctor can observe what they are like under analysis but never as they are outside of that influential setting.

There are equally clear cases of altering the object of observation in the physical sciences. If you put a big, warm thermometer in a little bit of cold water to measure the temperature, the water will heat up. The final reading on the thermometer will be a compromise between the original temperature of the water and that of the thermometer itself. It will be information about the interaction between the thermometer and the water, but it will not indicate the temperature the water would have been if we had left it alone.

The thermometer, of course, does not have to be big and the sample of water does not have to be small for one to affect the temperature of the other. I rigged the example this way just to make the effect more graphic. But the effect is always there to some degree or other when you stick a thermometer into a sample. In fact, some amount of physical change of the object takes place in almost every case of observation in physical science. An exception would be a case in which the water happens to be exactly the same temperature as the thermometer to begin with. In the majority of cases though, changes in the system are essential aspects of the observing processes. Observation requires the transfer of information through some interaction that changes the momentum and energy of both participants. We see things, for example, by reflected light bouncing off the object. This changes the momentum of the light and the object, and we get information on how the object has interacted with us but not how it would be on its own. Other changes in the object of observation are incidental but unavoidably caused by the proximity of the observer. Just by being in the neighborhood we raise the temperature of things just a little, and we cause subtle vibrations.

There are two possible remedies for this physical influence on observation, two ways to use the observation to nonetheless get information on the underlying reality. The first is to reduce the physical impact as much as possible. As an anthropologist learns to be unobtrusive, a physicist learns to use a small thermometer that is close to the estimated temperature of the specimen. But the benefits of these efforts have a limit and the physical effects of observation cannot be reduced to zero. One of the most noteworthy components of quantum mechanics explicitly describes the minimum of interaction that we can get away with in the act of observation. It is this principle in large part that motivates Bohr, and it is just this kind of thing that makes physics so epistemologically interesting.

The other option for dealing with the physical influence on observation is to carefully take note of it and conceptually reconstruct reality. If we are aware that there is a human contribution in what we see, and we know exactly what the contribution is, perhaps we can infer what the world's contribution must be by subtracting our own input from the final image. The example of the thermometer is a good one. Our knowledge of thermodynamics and of the specific composition of the particular thermometer allows us to know precisely how much heat the thermometer will contribute or absorb in any given situation. A good scientific thermometer will come from the manufacturer with data on what is called its water equivalent. With this data we can figure out just what effect the thermometer has on the temperature of the sample, and thus we can know what the temperature of the sample was before the measurement. In this way we are not trying to eliminate the effects of observing; we are trying to stay fully aware of the effects in order to abstract them out of the final report. This sounds great, but it remains to be seen whether this kind of reconstruction is possible in all cases, or indeed in any cases of foundational physics. It probably is not always possible to know in anthropology what influence you are having, and so it is not possible to systematically remove the effects of your influence from the description of the people. We will have to see how things go in physics.

NATURE AS IT IS

We have been working to clarify the notion of appearance; it is only fair that we do the same for reality. The idea is to make sure we know what we are talking about in the contrast between the two, and to have a focused idea on what to look for when we turn to physics.

In contrast to appearance, we expect reality not to change with our own changing perspective. The way things are will not be forced by our way of conceiving things nor by any caprice or planning in how we choose what to observe and what to make of it. The key is independence. Reality is what things are like when we are not looking or thinking about them. It is nature in itself, not nature for us.

We can talk in similar terms about the reality in the anthropological case. How people behave, what they think and value if never visited by anthropologists, that is the reality of the situation. A metaphysical realism would claim that there is a determinate way that these people are, independent of the anthropologists or any other inquisitive intruder. Epistemological realism would claim that furthermore, the anthropologist can figure out what that reality is like.

The realist claims about natural science and physics, both metaphysical and epistemological, are quite minimal at this point, without detail on what the reality is like. It is important that we keep them this way until the evidence in physics is in. For example, nothing has been said to indicate that events in the independent physical world are deterministic in the sense of being law-governed and non-random. So if a proof of indeterminism shows up in the physics, this does not necessarily tell against metaphysical realism. No particular property is essential to the existence of an independent world, and so even if we can prove that one particular property or another is essentially dependent on a human observer, this does not mean that all properties are like this. Proof that one or another property is missing from the real world does not indicate that all properties are missing and hence that there is no real, independent world. A stone, for example, is neither male nor female. The property of sex simply does not apply, just as the property of color does not apply to individual atoms. But just because one property is demonstrably indeterminate of a particular object, we surely would not conclude that the object lacks objective existence entirely. Keep this fallacy in mind when we come to quantum mechanics.

The epistemological realist position also has limitations that need to be respected. To claim that we can have knowledge about the way things are and not just the way they appear does not require that we can know *everything* about nature. To claim that we can know anything at all beyond the appearances is an epistemological realist position. The details of the position must be filled in to pinpoint exactly what we can know and how.

All of these cautionary notes against reading too much into the basic realist positions can be summarized by pointing out that discussions of realism should be focused on specific things and properties rather than on reality in general. Do not ask whether reality is independent of the observer; ask instead whether color, or position, or mass is independent of the observer. Any available evidence to answer the question will be about these kinds of particulars.

Another way to put this advice is to ask about reality as an adjective rather than as a noun. Ask, that is, whether color is real in the sense of having a definitive existence independent of an observer. But avoid questions about all of reality, or, as people sometimes put it, Reality. Philosophers have a peculiar habit of elevating adjectives to nouns, of talking about the redness of an apple, for example, as if there is something else that exists in addition to the apple. In our case of analyzing what physics allows us to know about nature, this nouning will only bring on confusion that cannot be resolved by the evidence. With this in mind, we will always focus the issue of realism onto specific things or properties, and never take on the whole world, or even the whole World.

PHILOSOPHICAL ARGUMENTS ABOUT REALISM

The discussion of realism in this chapter is entirely general and abstract, about the theories in general and observations in general and how they are generally related. It is philosophy, and in so far as it comes before the evidence in the physics, it is intended to be only suggestive of how the issues can be approached. These are the hypotheses to be tested and refined or perhaps rejected in light of the evidence. Here we are only outlining the basic concepts that will guide the exposition of the physics and tell us what to look for. We are also clarifying the challenges for an attitude such as realism, and the alternatives. Since it is advisable to know the challenges before digging into the task, it will be worth making a quick survey of philosophical discussions of realism.

Most philosophical arguments about scientific realism focus on the question of whether there is good reason to believe that scientific theories are true. At this philosophical level of abstraction, suggestions are made that perhaps we are justified only in believing what the science claims about things we can verify by observation, and the rest, the more theoretical claims about unobservables, are to be treated with skepticism. Why we can believe the claims about unobservables (or why we cannot), or why we are justified in believing a theory to be true (or why we should regard it as a useful, though not necessarily true, way of thinking), are central concerns of the philosophical debate. But they are not exactly our concerns. Our question is *not*, should we believe the theory of relativity or quantum mechanics is true? Rather, our question is, *given* the claims made by relativity and quantum mechanics, what are the implications for our ability to know about nature beyond how it appears to us? The idea is not to use a philosophical account of realism to illuminate the science; it is just the opposite, to use the science, the details of the theoretical claims, to illuminate the issues of realism. We are not out to assess the proofs of these theories but to see whether these theories reinforce or alleviate any of the philosophical concerns over realism or anti-realism.

We have already seen a good philosophical reason to seriously consider anti-realism and to side with Bohr, or at least my paraphrase of Bohr. Observation, the appearance of nature, is partly a product of our own intervention. Observation, as philosophers often put it, is theory-laden. How then can observation be used to test theories about nature in an objective way? Observation is also laden with the physical influence of our perspective and the interaction that is the act of observing. We are always in the picture, in both a conceptual and a physical way. How then can we possibly know what nature is like without us?

There is another forceful reason to take anti-realism seriously, often called underdetermination. The idea is that no amount of evidence has sufficient information to determine which of the many possible theories about nature is true. Appearance underdetermines reality, and so we cannot use appearance to prove what reality is like. For all the things we see happening in nature there will always be more than one explanation, more than one interpretation of what the evidence means about the unseen mechanisms behind it. The evidence cannot indicate which of the alternative theories is true.

The logic of this underdetermination argument is very neat and very persuasive. Testing a theoretical claim in science is generally based on its success in explaining phenomena that have been observed, or in predicting phenomena that are observable in the future. Prediction and explanation have the same basic form: *If* the theory is true, *then* such and such phenomenon will be observed. The theory is tested by looking to see if the phenomenon does (or has) happened. But even if it does happen, that is, even if the prediction or explanation is a success, we cannot conclude that the theory is true. Surely there is an alternative theory that describes reality differently but makes exactly the same predictions and succeeds on the same explanations as the theory we had in mind. Lack of imagination and complacency with our originally successful theory may prevent us from looking for the alternative, but it is there, equally supported by the evidence.

The point is that the appearance of nature does not single out a particular theory of the reality of nature. Different putative realities can appear the same, and so we cannot use the appearance to prove which *one* of these possibilities is really how nature is. Prudence and epistemic responsibility, the essence of scientific method, would have us believe only the scientific claims about what can be observed. Theories, the claims about what cannot be observed, can perhaps be used as helpful models, ways to organize our thoughts but not the objects of belief. Theories are useful, but not true. Nature acts *as if* our theories about it are true, but that does not mean that nature *is* as these theories say.

There are also philosophical arguments in favor of scientific realism. One of the most common is presented under the heading of inference to the best explanation. It points out that the reasoning used in science to support claims about the reality of nature is an extension of a very common-sensical and natural pattern of thought we use all the time in life.

We infer that what best explains the things we see is probably true. For example, if the water drains very slowly from the bathtub, and then one day stops draining altogether, I infer that somewhere in the dark, inaccessible, unobservable reaches of the pipe is a clot of hair and soap. This would explain a backup of water. There are, of course, lots of other explanations. Maybe someone poured cement down the drain. Maybe aliens have taken up residence in the pipes. There are these and other explanations, but the hair-and-soap clog is the *best* explanation, given other things we know about bathtubs and plumbing. We must admit that the alternatives *might* be true, but the best explanation is the one most *probably* true. In this and other mundane cases we infer from the appearance, the evidence of the stagnant water, to a justified (though not dead certain) claim about the reality. This inference to the best explanation is the way of common sense and the way of science, since the business of science is to identify and articulate the best explanations for the phenomena of experience.

The credibility of the inference to the best explanation, and its value to the cause of realism, rely on the clarification of two key points. Each is vulnerable to challenge by the anti-realist. First of all, what makes one explanation better than another? And in particular, what counts as the *best* explanation? Is it the simplest? Simplicity itself is an ambiguous concept and its assessment is often controversial. Which explanation of the diversity of life as we observe it now is simpler, the theory of evolution or the creationist account? As an alternative, or a complement, to simplicity, perhaps the best explanation is the one most compatible with other things we know about nature, other theories. This is why I ruled out the theory about aliens in the plumbing. This counsel of conservatism, to accept the explanation that makes the most sense given what we already know, is surely good advice, but it amounts to justifying one theory that the anti-realist would say is suspect, by appeal to other theories that the anti-realist would say are suspect. Justifying new ideas about reality on the basis of old ideas about reality might just be perpetuating old misconceptions. Furthermore, there are times in the history of science when a new idea was accepted despite its incompatibility with firm convictions about reality. If we never accepted an explanation, an account of reality, that defied the entrenched theories, we would still be claiming that the earth is the center of the universe and that everything is made up of the four elements, earth, air, fire, and water. The point is that sometimes the explanation that contradicts other things we believe is the right one, so either there is some other important criterion for what counts as the *best* explanation, or being the best explanation is not indicative of being true.

And what does explanation, even the best explanation, have to do with truth? This is the second loose end in the argument that appeals to inference to the best explanation. The accomplishment of explanation, after all, is a psychological accomplishment. An explanation is a claim that satisfies our curiosity, a claim that gives our questioning a rest, at least momentarily. A false claim could do this as well as a true claim. There is a burden of proof still to be met by the realist, and that is to establish the link between explanation and truth. Inference to the best explanation *seems* right. It is intuitive, but in science (and philosophy), we demand more than intuition. We need proof.

Some of the realists' responses to these challenges come in the form of refining and specializing the claims of realism. There is, for example, an explicitly causal form of realism that is based on inference to the best *causal* explanation. Some explanations cite the cause of the phenomenon to be explained, others do not. A non-causal explanation might make sense of a phenomenon by simply fitting it into a larger category of similar events. The causal explanations, according to this realist argument, have more than a psychological value. They have an epistemic value; they are indicative of the truth because something must be causing the phenomena we observe. Evidence that x *explains* a phenomenon is not in itself reason to believe in x, but evidence that x *causes* the phenomenon is reason to believe in x. Explanations are in the mind, but causes are real.

There is a related version of realism that contrasts realism about theories with realism about things. Entity realism concedes to the skeptic that scientific theories about unobservables cannot be proven by the evidence, but maintains that something, some entity, must be responsible for causing the observations and interactions we experience. We may be wrong on the theoretical details, but we know that things like electrons and photons, things we cannot observe, must be real. This minimal inference from appearance to reality at least is warranted.

Anti-realism is sometimes summarized by its detractors as relying on a miracle. It would be a miracle for a theory to get all the explanations and predictions right and yet still be false. How else do you explain the empirical success of a theory other than by its being true? The realist, in other words, seems to have a plausible explanation for why some theories do so well when compared to the evidence. It is because they are true. The truth of the theory is offered as the best explanation of its success, or at least as a better explanation than anything the anti-realist has to offer. This sounds good for the realist until you notice the circularity in the argument. At the heart of realism is the dependence on inference to the best explanation, and here we are arguing for realism by saying that it is the best explanation of what is observed as a theory's explanatory and predictive success. This is a case of using inference to the best explanation as the proof that inference to the best explanation is a legitimate form of argument.

And so the philosophical debate continues. It is important to remember that our concern here is not so much with the influence of this debate on modern physics as with the influence of modern physics on the issue of realism. The question is whether the details of relativity and quantum mechanics support realism or anti-realism, or neither. Does the physics suggest more reasons, beyond those in the philosophical arguments, for or against realism or anti-realism? Does the physics reinforce any reasons for anti-realism already mentioned, underdetermination perhaps, or the human influence on observation? Does the physics respond to any of these challenges to claims to know reality, perhaps in some particular cases?

The physics, we will find, is rich in relevance for realism. The special theory of relativity, though its name might suggest a total capitulation to dependence of appearance on the circumstances of the observer, in fact shows explicitly what is and what is not observer-dependent. And by carefully describing the dependence, the theory is a guide to dealing with it, to understanding our own contribution to the appearance so we can learn about nature's contribution. The general theory of relativity will reveal a clear case of underdetermination of theory by evidence. The physics in this case will clarify a limit between what we can know and what we cannot. Quantum mechanics will be explicit in describing the human complicity in the act of observation and so will also provide information to distinguish what we can know from what we cannot. The results of the physics will not be a general endorsement of either realism or anti-realism. It will force us to deal with the issue on a case-by-case basis. Some things we can know, some we can't. And for some important examples we will find out which is which.

RETURN TO BOHR

> It is wrong to think that the task of physics is to find out how
> nature *is*. Physics concerns what we can say about nature.

Bohr distinguishes between two possible accomplishments of physics. We might hope to describe ("to find out") nature itself, as it really is, or we might only be able to describe what we know ("what we can say") about nature. In light of the previous analysis of observation and the nature of appearance, it is worth adding a third alternative to this list. We might be able to describe only how we interact, both conceptually and physically, with nature.

Between the first accomplishment and the third is the contrast we usually associate with the distinction between objective and subjective knowledge. Between the first and the second there is no real distinction at all. The second is, in fact, redundant, claiming that we can describe only what we can say. This is certainly a general limitation on knowledge and the predicament is in no way distinctive to physics.

The accomplishment of objectivity in science is to get from the third kind of information to the first, to use knowledge of our interaction with nature to know about nature itself. The accomplishment of objectivity is not in escaping our own point of view (because that is impossible) but in understanding our point of view. The goal is then to use the composite appearance, composed as it is with information from the world and from ourselves, to know about the world's contribution. This is the proper compromise between taking observation at face value, at one extreme, and forcing the hand of observation by overbearing theories, on the other. In the former case, with no conceptual influence, observation will be meaningless and only haphazardly reliable. In the latter, conceptualized observation may lose the information from nature and serve to test theories only in a circular, self-serving way. The compromise is in explicitly recognizing both the need and the effect of subjective influence on observation and thereby recognizing the effect of nature as well.

Even in Bohr's assessment of the appropriate task of physics there is hope of discovering how nature *is*. "Physics concerns what we *can* say about nature." So apparently there are things we cannot say. The description of nature is not a free-for-all; it is constrained, most likely by nature itself. This is exactly the sort of link we want to exploit between appearance and reality.

The plan now is to turn to physics to see just what we can say about nature and what we cannot. The driving question is whether we can responsibly claim to know about aspects of nature that cannot be observed. Can we know how things are, based on how they appear?

Chapter 3
THE SPECIAL THEORY
OF RELATIVITY

THE PRINCIPLE OF RELATIVITY

The laws of physics apply universally. This is almost in the definition of physics, the science of the basic properties and laws common to everything. Gravity, conservation of energy, action and reaction, these are true of organic things and inorganic, large and small, Republicans and Democrats.

The laws of physics apply not just to all kinds of things but to things in all kinds of circumstances. Matter and energy have the same characteristics and are governed by the same principles here on the earth as on the moon or on the sun or a distant galaxy or anywhere in between. Achieving this kind of universality was part of Isaac Newton's accomplishment, the part about the apple and the moon. Recognizing that a force was pulling the apple to the earth and causing it to fall, he figured that exactly the same force, gravity, affected the moon and held it in orbit around the earth. The important discovery is not just of gravity, but of *universal* gravitation.

If the laws of physics are exactly the same everywhere in the universe, then you cannot use the laws of physics by themselves to tell where in the universe you are. There is no experiment with a law of physics, the relation between force and acceleration, for example, that will turn out differently in different places. You cannot use the laws of physics as you might use language to figure out which country you are in. Language differs from place to place, and if everyone is speaking Chinese, chances are you are not in Mexico. If there was only one universal language, as there is only one set of laws of physics, then language would be of no help in distinguishing places.

The laws of physics are the same in all different places; they are also the same in all different circumstances of motion. In other words, the laws that apply in a classroom that is at rest, not moving on the earth, apply as well in a moving train or airplane. This is really convenient. You do not have to learn different versions of physics for different circumstances of motion you might experience. Nor do we need different editions of the textbooks, one that describes how things work in a lab that is standing still, another on how things work in a compartment going 10 miles per hour, another at 15 miles per hour, and so on. This universal aspect of the laws of physics has the same corollary as before. You cannot tell whether or not you are moving by determining which laws of physics apply in your circumstances. If you find that an object accelerates in proportion to the external force, it does not mean you are not moving, since this relation between force and motion is true in a moving train as well. Toss a bagel to your friend sitting across the compartment and it will follow the same parabolic trajectory whether the train is standing still at a station or racing down the track at a 100 miles per hour.

There is something of a problem though when the train is accelerating, that is, when it is speeding up, slowing down, or turning a corner. In these circumstances, suitcases fly off the overhead rack, passengers are pressed against the wall, and the tossed bagel veers recklessly into the lap of the taciturn businessman sitting next to your friend. These seem to be events in which objects accelerate with no apparent forces acting on them. In other words, the law that acceleration is proportional to force, $F = ma$, does not seem to be working and we can use this violation of the law to distinguish accelerating circumstances from non-accelerating. You *can* tell, by doing a simple experiment, whether or not the train is accelerating. You can feel it.

The difference between accelerating and non-accelerating circumstances of doing physics is important. It is one of the keys to understanding relativity, and we will return to it with all of the detail and depth that it deserves. But for now we will deal only with physics in non-accelerating circumstances. For the moment, let the train go as fast as you want but do not let it turn corners, speed up, or slow down. This restriction is what is specialized about the special theory of relativity.

The discussion so far about the universal application of the laws of physics can be neatly summarized with the concept of a reference frame. The platform of a train station is one reference frame. A train moving by is another. To fill in the details of the platform reference frame, we choose one point, a benchmark from which to measure the distance to any event of interest. This point is called the origin. At the origin we also choose directional axes to express not only how far away an event is from the origin but also in which direction. A convenient choice would be one direction along the tracks, another that is the vertical, up-down direction, and a third that is the horizontal line perpendicular to the tracks. In a three-dimensional world we need three directional axes. Together, the origin, the directional axes, and a unit of measuring distance, a unit like feet or miles or meters, constitute a coordinate system. The origin and axes remain fixed in the reference frame. In the platform frame, the origin is *always* in the same place and we always specify direction in terms of along the tracks, up-down, or perpendicular to the tracks. In general, a coordinate system and its specified state of rest is a reference frame.

The train reference frame has a coordinate system fixed to the train, and the whole thing is moving with respect to the platform. The businessman who is now asleep after finishing his bagel is not moving with respect to the train reference frame. He is moving at 100 miles per hour (or whatever the train's speed) in the direction along the tracks, with respect to the platform reference frame. A car speeding down the road that runs parallel to the tracks is, if its speed and direction match the train's, at rest in the train frame. It, like the napping tycoon, is moving with respect to the platform frame.

It is important to note that a reference frame does not require an observer. Nobody needs to be on the platform or in the train for us to talk about the platform frame or the train frame. A runaway, unoccupied train makes for a fine reference frame.

Now we can summarize the idea about physics being universal. In this chapter, that is, in the special theory of relativity, we will specialize to reference frames that are not accelerating. These are called inertial reference frames. While the train is taking a turn or slowing down or speeding up, the train frame is non-inertial, and we will deal with that later. For now though, from our own experience and from the earlier account of the universal nature of physics, we can say that the laws of physics are the same in all inertial reference frames. There is one set of laws, including the law $F = ma$ we have been using as an example, for all inertial reference frames. No inertial reference frame is singled out as special, as the one in which our physics is true. At this point though we have identified a special *class* of reference frames, namely those that are inertial. Within that class, physics experiments cannot tell us which frame in particular we are in. When you wake up on the train and the window shades are closed, you cannot tell if the train is moving or not. Similarly, when the adjacent train at the station begins to move relative to you, it is hard to tell who is moving away from the platform, you or they.

In this talk of universal laws of physics we have already stated the **Principle of Relativity**, a conceptual pillar of the special theory of relativity:

> The laws of physics are the same in all inertial reference frames.

Notice how inappropriately this principle is named. The word "relativity" here could not be more misleading. The principle says that the laws are *not* relative to a reference frame. It says that the world as described by physics does not change with the change of perspective from one reference frame to another. "Relativity" is not a good name for this principle, but we are stuck with it.

The principle of relativity is a foundation of Einstein's special theory of relativity, but the principle was not invented by Einstein. It has been an important guide to doing physics at least since Galileo. The special theory of relativity though represents an unprecedented fidelity and consistent application of the principle.

SPACE AND TIME

Before using the principle of relativity to develop the details of the special theory of relativity, we should get back to the distinction between inertial and non-inertial reference frames. An inertial frame, recall, is one that is stationary (like the platform frame) or moving in a straight line at a constant speed (like the train frame). But a larger view of things reveals that the earth itself is in motion. The platform, fixed as it is on the earth, is spinning around with the earth and orbiting the sun with the earth. And the sun itself moves, swirling in the whirlpool that is the Milky Way galaxy. So if the platform frame is inertial by virtue of being stationary, we have to ask, stationary with respect to what? The important question here is not which frames are inertial, but more generally of the criterion to distinguish inertial from non-inertial frames. We do not want a list; we want a method. An inertial frame is stationary or in uniform motion with respect to what? The earth cannot be the standard, because that would undermine the universal status of the laws of physics, making them all refer to our planet. Besides, the earth moves through space.

But what is space? Maybe space is the basic reference of inertial frames, in the sense that "inertial" means at rest or in uniform motion through space itself. Asking about the nature of space is probably a violation of Bohr's advice about the appropriate task of physics, but we can give it a try and see where it leads.

There are two possibilities for the nature of space. It could be something in itself with its own properties independent of the objects and events in space, or it could be nothing more than the relations between the things. The former description is often called the substantival model of space. The latter is the relational model.

In the substantival description, space itself is a substance in which the objects of the universe exist and move. This does not mean that space is something you can cut with a knife or sell by the jar. To call space a substance is just to say that, if asked to list everything that exists in the universe, somewhere on the list you must mention space itself. In fact you might want to start with space: first, there is the space in which everything else is situated, the container, so to speak. It is a unique substance with no mass, no color, no smell, no perceptible properties at all. The idea of empty space makes sense if space is a thing in itself, and we can imagine a universe completely without matter. Concepts of position and motion would still make sense because they would be position in space itself.

A similarly substantival description of time is possible. If it makes sense to even speculate that all physical processes, all motion in the universe, stop for, say, ten seconds, then time itself must have an existence independent of physical things and events. If nothing happens but the passage of time, then time, like substantival space, is on the list of things that make up the universe.

Contrast this account of the nature of time with a more relational description in which time simply *is* the changes in physical objects. In the case where nothing happens, no time passes. It makes no sense, in this model, to say that all processes stopped for ten seconds, because "ten seconds" simply means some number of vibrations of a particular atom (or whatever standard we choose). With no vibrations there are no seconds, no passage of time.

This account of time is like a relational model of space. On this account of space, there could be no such thing as a universe with nothing in it but empty space. Without physical objects as reference, there is no space. In the actual universe, only physical things exist. These things bear certain relation to each other, like Jupiter is bigger than Pluto, New York is closer to me at the moment than is the sun, the moon is more massive than I am, the train is moving with respect to the platform, and so on. Some of these relations are spatial. To-the-left-of, are-one-mile-apart, and the like are spatial relations. But without real objects to be a mile apart, for example, there is nothing. Space is a collection of properties of things; it is not a thing in itself. And the things we are talking about have spatial properties only in relation to other things, not to any ubiquitous, ethereal, substantival thing that is space itself. To say that the train is moving means it is moving with respect to the platform. To say that the earth or the sun or the galaxy is moving means that it is changing position relative to some other object. It often goes without saying just what other object is the point of reference for position or motion. When I say the train is moving, it is simply understood that it is moving with respect to the platform. But it is always an incomplete description to say that something is moving or sitting still. Moving in relation to what? Saying simply that the earth is moving is as meaningless as saying that the earth is bigger. Bigger than what?

This is the relational model of space. The important challenge raised by discussing the relational and substantival models is in deciding which is true. Is space really only relational, or is it really substantival? This is a question about the way nature *is*, and either answer would be idle speculation unless there is some observational evidence one way or the other. This is not a popularity contest; this is science, and we need good reason to believe descriptive claims about nature.

We are looking for observational differences between the two models of space. On neither model is space itself visible or perceptible in any way, so any observational evidence that distinguishes one model from the other will have to be indirect. We are looking for the effects of space being relative or substantival. If the evidence is neutral, if nature is equally as if space is relational as it is as if space is substantival, then the burden of proof seems to weigh more heavily on the substantival account. Substantival space is extra metaphysical baggage that we need good reason to carry.

Gottfried Leibniz, early in the eighteenth century, used this kind of argument about parsimony to conclude that the substantival model is untenable. There is no evidence at all, he claimed, for a unique reference frame of space itself. We measure position and motion of objects only in reference to other things. If everything in the universe were moved, say, a foot to the left, there would be no discernible difference. All measurements, all observations would be unchanged. Similarly, if everything were moving uniformly to the left we could not tell, again because all objects and hence all points of reference would remain at the same separation and the same relative motion. Thus, substantival space cannot be detected either by direct observation or by indirect evidence in measuring the positions of objects or their uniform motion. An entity that neither serves a purpose in explaining the phenomena we do see, nor is observable itself, has no legitimate role in science.

Isaac Newton, in defense of substantival space, endorsed the same standard of scientific proof as used by Leibniz. A scientific claim is acceptable only if it has clear, observational evidence. Newton had to admit that there was no way to detect an object's position in the reference frame of substantival space. Furthermore, there is no way to detect uniform, that is, inertial motion with respect to space itself. But non-inertial motion is different, and Newton claimed that we *can* detect acceleration with respect to space itself.

Here is a version of the simple thought-experiment Newton used to show that space is more than a set of relations between objects; space is a thing in itself and we can detect the unique reference frame of the substantival space. Partly fill a regular bucket with water and put the bucket on the turntable of your old record player. Notice that the surface of the water is flat. Now turn on the turntable so the bucket is spinning around the vertical axis through its center. At first the bucket spins but the water does not, and the surface of the water remains flat. But soon the bucket begins to pull the water around with it and, as the water itself spins, the surface takes on a concave shape. Water creeps up the sides of the bucket, perhaps even spilling over the edge, as long as the water is spinning. Even when you turn off the turntable and the bucket stops spinning, the surface of the water will be concave for the little while the water itself continues to spin.

The crucial observation in this experiment is that there is a clear and distinctly observable difference between spinning water and not-spinning water. The surface of spinning water is concave; the surface of not-spinning water is flat. But here is the important question: Spinning with respect to what? Newton claimed that the only explanation for the concave shape of the water's surface is that the water is spinning with respect to space itself. It is not spinning with respect to us that pulls the water up the sides of the bucket, nor is it spinning with respect to the room. You could put the whole room on a merry-go-round, a huge turntable, and still the water would be concave when it and the whole room were spinning.

Newton's point here is that non-inertial motion like spinning does not require another object of reference to complete the concept. To say that the water is spinning is to give a complete description because we know it means spinning with respect to space. In a universe with no objects other than the bucket of water we could tell if it was spinning by seeing whether the surface was flat or concave. Substantival space, in other words, has an observational effect on non-inertial motion. A real effect, namely the changed shape of the water, requires a real cause, namely a real, substantival space.

Talking about what we would see in a lonely universe with only a bucket of water is a handy, and safe, heuristic exercise. We will never actually make this observation, and its outcome is not beyond a reasonable doubt. Ernst Mach, at the end of the nineteenth century, raised such doubts and reopened the case of Newton's bucket to reconsider the relational model of space. Mach, it is worth noting, was cited by Einstein as an important inspiration for the theory of relativity. Mach proposed an explanation for the concave surface of water that did not involve any substantival space. He claimed to account for the spinning bucket demonstration strictly in terms of the simpler, less speculative relational space. Mach's proposal is that the water is pulled up the sides of the bucket by distant massive objects, the stars and galaxies in the universe. Somehow, when the water and distant stars are in relative motion there is an extra force, in addition to gravity, acting on the water. Thus, the water is concave when it spins with respect to the massive objects in the universe. Space itself has nothing to do with it. In a universe without these massive objects, if there is only a bucket of water, the water would always be flat. Not only could we not tell if it was spinning or not, it does not even make sense to say it is spinning or not, since there is no point of reference, no other object, by which to define motion. In a relational universe with only one object, motion and position are vacuous concepts.

By Mach's account of the bucket on the record turntable, we must say that switching the record player on causes the bucket to spin with respect to the massive objects in the universe. You could look at it in two ways. Flipping the switch gets the bucket turning as the stars and galaxies hold still. Alternatively, the bucket sits still while all the cosmos is caused to orbit at dizzying speeds. Only motion relative to some other object is observable or has observable effects. This goes for inertial motion and for non-inertial motion.

We do not know whether Newton or Mach is right about the surface of the water in a bucket alone in the universe. As thought experiments go, this one is not particularly helpful. Furthermore, Mach never worked out the details on how this newly proposed dynamic force between distant stars and the water in the bucket is supposed to work. Nobody else has either, and this remains an impediment to Mach's case against substantival space.

As it stands then, neither the substantival nor the relational argument is conclusive. Maybe Bohr is right and we should not even be asking questions about the nature of space. But I'm not ready to give up, and the appropriate thing to do next is to see what, if anything, contemporary physics, namely the theory of relativity, can say on the issue.

RELATIVE AND ABSOLUTE PROPERTIES

The substantival/relational debate pivots more directly on the concept of motion than on space itself. Since we all agree that there is no chance of observing space, its nature must be finessed by observation and analysis of how observable things are situated and how they move. In the substantival model we can talk about motion or location pure and simple without having to refer to other objects as benchmarks. There is a cosmic reference frame, a unique absolute standard of rest. In the relational model though, motion makes sense only with respect to other objects. A report that we are moving at 2 meters per second, or a bird is flying at 10 m/s is incomplete unless the reference frame is mentioned or previously understood.

It is worth pointing out that motion is a *property* of a thing; it is not a thing in itself. This is an important distinction worth keeping track of. So is the distinction between properties that are relative and properties that are absolute. An absolute property is one that is intrinsic to the object or event, intrinsic in the sense that it is independent of the external circumstances. An absolute property does not depend on anything other than the object it is a property of. It is not in relation or comparison to anything else. A relative property, on the other hand, is one that depends on the circumstances. The property is not fully realized unless both the object and some particular circumstances are identified. Do I have the property of being taller-than? In some circumstances I do, as when I am compared to Quasimodo (as often happens), but in other circumstances I do not, when I am compared to Michael Jordan (ditto). Taller-than is a relative property. Do I have two eyes? Yes, period. This requires no comparison, no account of my particular circumstances. I have two eyes in any case. Having two eyes is an absolute property. It is invariant in that it does not change, either as I move around or as someone observing me moves around.

With this distinction the substantival/relational differences can be summarized. In the substantival account of space there is a concept of absolute motion. In the relational, all motion is relative. Newton would claim that the water with a concave surface is spinning, period. Mach's response is that it is spinning relative to the distant stars, just as they are spinning relative to it. To a bug on the turntable, the water is not spinning. Spinning or not, according to Mach, depends on your point of comparison, your reference frame.

We need an even easier example to sort this out and to get comfortable with the absolute/relative distinction. Consider an ordinary object like a pencil that is at rest in the room, sitting on a table. Of all the properties of the pencil we can group them definitively into those that are relative and those that are absolute. Focus on those properties that might reveal the nature of space and time, so ignore things like the pencil's smell, the hardness of the lead, and the like. Common sense would say that the position of the pencil is relative, while its length is absolute. Please note that common sense may suffer some revision when we turn on the physics, but for now it will do.

Position is relative because any indicators of where the pencil is, to-the-right, 6 meters away, and the like, will depend on our choice of a point of reference. Specifying its location requires a reference frame with a coordinate system, so position is relative to the reference frame. We can imagine two different reference frames in the room, call them K and K', that differ only by their choice of origin. I use system K with its origin where I am sitting. You use K' with its origin where you are sitting.

Notice in Figure 3.1 that reference frames K and K' use the same directional axes. Each has a third axis, z in the K frame and z' in the K' frame, that is horizontal and points directly out of the plane of the paper. But it is too hard to draw this axis, and it is not crucial to the discussion, so I have left it out.

The only difference between K and K' is that their origins are a horizontal distance d apart. Say that K' is a distance d to the right of K, or that K is a distance d to the left of K', and you say the same thing.

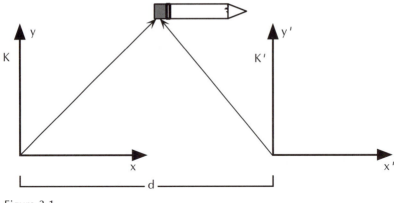

Figure 3.1

Where is the pencil? Or, to be more focused on a single point, where is the eraser of the pencil? From the K perspective it is some distance up (along y) and some distance to the right (along x). From the K′ perspective it is the same distance up but some distance *to the left*. Position is relative to the benchmark of measurement and we cannot just ask where the eraser is. We have to ask where is it with respect to a particular reference frame.

The length of the pencil is different. It is, at least on this intuitive level of analysis, an absolute property. The question here is not of the appearance of length as viewed from each origin. A K observer can certainly move around in the K frame, just as a restless passenger can walk around in the train. Only the point of reference, the origin and axes are fixed. So the length of the pencil should be measured by each observer as carefully as possible, not being fooled by foreshortening or the shrinking effects of a distant view. Since the *difference* between the position of the point and the position of the eraser will be the same as measured from either, indeed from any, reference frame, the length is absolute. Again though, this is just common sense talking. The concepts of what it is to be absolute and what it is to be relative will never change, but whether we call a particular property one or the other might change as we look more closely at the details.

Is the pencil moving? Both K and K′ say no; the pencil is stationary. Their respective values of position are different, but neither changes over time. But "stationary" means that something is moving at the same speed and same direction as the reference frame, as the pencil is moving along on the earth with the two frames in the room. For another reference frame, K″, that is moving past the room with a uniform speed s, going to the left, the pencil is moving. It is moving to the K″ right with a speed s. Its position in the K″ coordinate system is changing over time, though the difference between the point of the pencil and the eraser stays the same. Thus, length is an absolute property, but uniform, inertial motion is relative. The former is invariant from reference frame to reference frame, while the latter changes.

The point of the spinning bucket experiment was to show that non-inertial motion is an absolute property. Any reference frame will measure the water's surface to be concave, when it is, or flat, when it is. Flat or curved, like the length of the pencil, will be invariant frame to frame, so all frames agree on whether the water is spinning or not. Rotation, a kind of non-inertial motion, is absolute. Newton then used this conclusion about motion to support his case for substantival space. Space with its own independent existence and a unique reference frame is necessary to distinguish between inertial and non-inertial motion, a distinction that has observable consequences.

Getting back to the pencil, we should be able to classify each of its properties as either absolute or relative. What about time duration? This is not a property of an object but of a happening. As a specific example let us say a bug has decided to eat the eraser of the pencil. One property of this process is how much time it takes, its temporal length, analogous to the spatial length of the pencil. Is this an absolute or a relative property? My intuition says absolute. Give all the observers, K, K', and K", synchronized, identical clocks, as we gave them identical length-measuring sticks, and they will all measure the same time duration. Neither spatial displacement, as between K' and K, nor uniform motion, as between K" and K, will change the time duration. So says common sense.

It would be nice to have a visual diagram of this situation as we did for length and position measurements. We could do this with little icons of clocks to mark the time of the bug's first bite (call this event A) and the time of the bug's final bite (event B). But we need to show the circumstances of the reference frames as well. The previous diagrams (Figure 3.1) showed two spatial axes. These space-space diagrams leave no room to show the duration of time. We will use instead a space-time diagram where the idle space axis (the vertical y axis revealed no differences between K, K', and K" measurements, anyway) is replaced by an axis for time.

There is an important difference between the space-space diagram of Figure 3.1 and the space-time diagram of Figure 3.2. The space-time diagram is an abstract, mathematical representation of events. The time axis is not a physical, straight vertical line. Time is not another physical dimension; it is just another variable we want to keep track of and we can do it in a convenient, visual way with a time axis. The space-space diagram is a more literal, physical representation of things. The y axis is, just as it is drawn, a physical, straight, up-and-down direction. You could build the space-space diagram out of sticks and place the pencil at its position in the physical coordinate system. You could take a photograph of it. But you cannot build a space-time diagram in a way that the time axis really points in the direction of the passage of time. It is just a mathematical representation of the situation, as the word "apple" is a linguistic representation of a kind of fruit. The representation does not look anything like the real, physical thing.

On the space-time diagram in Figure 3.2, both events A and B, the first bite and the last, happen at the same place according to the room reference frame K. So the value of x, the distance from the origin, is the same for A and B. But since B happens after A, the value of t, the time for B, is greater than that of A. B is directly above A on the space-time diagram. The line between A and B is the collection of events at the eraser between A and B, namely the second bite, the third, and so on.

A stationary object like the eraser occupies the same point in space over successive points in time. It shows up as a vertical line on the space-time diagram. An object moving to the right, the pencil sliding across the table, say, occupies points in space that get further to the right along the x-axis as time goes on. This sequence of events is represented as a straight line that angles up (because time is passing) and to the right (because the object's point in space is changing to the right). The trajectory through space and time is called a worldline. An object moving faster to the right than the eraser would have a worldline that is tipped more to the right, that

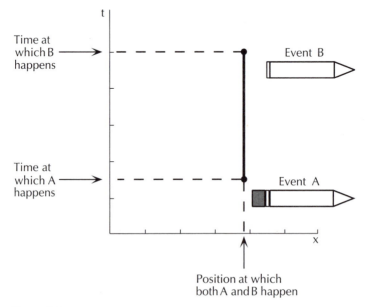

Figure 3.2

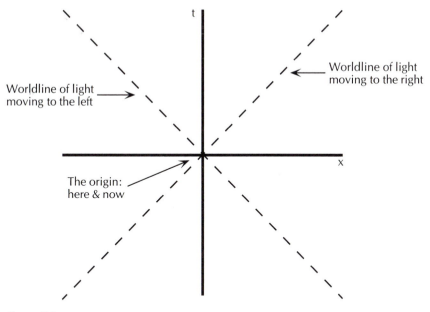

Figure 3.3

is, at a lower angle, since, during any interval of time it changes its position in space by a greater amount.

For precision in comparing things that are faster or slower we need to calibrate the spatial x-axis and the temporal t-axis. We can choose any units we want. The x-axis can be marked off so that each hash-mark represents a meter or an inch or whatever. Time axis hash-marks can be in seconds or hours or years. But we will soon be making a big deal about light and the speed of light, which is 3.0×10^8 meters per second. For this reason we will calibrate all our space-time diagrams such that something moving at the speed of light is represented by a worldline at 45°. So if each unit on the time axis is 1 second, each unit on the space axis must be 3.0×10^8 meters.

A space-time diagram that is calibrated in this way and that has the 45° line, that is, the worldline, a flash of light would take if shined from the origin, is a Minkowski diagram. Figure 3.3 is a Minkowski diagram waiting for us to draw in the worldline of some object or some sequence of events. Note that the dashed lines are worldlines of light shined horizontally along the x-axis. The worldline is tipped up at 45° because as the light passes through space it also passes through time. Its trajectory through space is horizontal, but its trajectory through spacetime, that is, its worldline, is at 45°.

Also note that the Minkowski diagram can be generalized to include the other two spatial dimensions, y and z. One of them can be drawn perpendicular to the paper. The third cannot be drawn, but we should have no problem with the concept. This is, after all, an abstract representation.

Figure 3.4 shows worldlines of a few typical phenomena as observed from the K reference frame. One is of an object at rest, at a point to the right of the origin. The other is of an object that approached the origin from the right, stopped at the origin and stayed for a while, and then went back whence it came but at a faster speed than it approached.

THE FOUNDATIONS OF THE SPECIAL THEORY OF RELATIVITY

Minkowski diagrams will be useful for determining which properties of things are relative and which are absolute. All we have to do is draw two Minkowski diagrams for two distinct inertial reference frames and see if the property in question is the same in both.

The laws of physics can be thought of as properties of things. Surely they are not things themselves; rather they are complex characteristics of

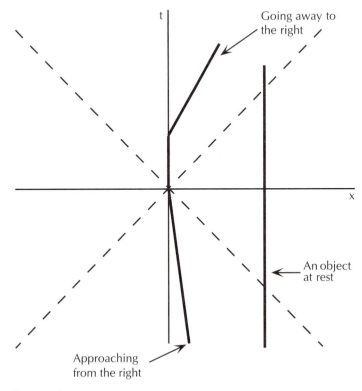

Figure 3.4

how things behave. The law F = ma, for example, says that objects have the property that their acceleration is proportional to the external forces acting on them. The laws of physics being properties of things, we can ask, are they relative or absolute properties? The principle of relativity is exactly the claim that the laws of physics are absolute. So, some individual properties of things, properties like position and speed, are relative and may change from frame to frame, but in every frame they will still be *related* in the same way. Any changes in one property in a different frame will be compensated by changes in other properties to which it is related in a law of physics. So the law of conservation of momentum, for example, has exactly the same form in one inertial reference frame as in another, though the values of the individual parameters like speed might be different.

The laws of physics describe more than just the mechanical phenomena in nature, the interactions of physical objects that have mass, size, position, and speed. There are also laws of electricity and magnetism, and these were initially somewhat recalcitrant to the principle of relativity. Figuring out how to reconcile the laws of electricity and magnetism with the principle of relativity, that is, figuring out how to make these laws invariant in all inertial reference frames, was the key to the special theory of relativity. It brought on a new understanding of space and time. This was one of Einstein's achievements, the one that makes the special theory of relativity his theory.

A stationary electric charge exerts an electric force on other electric charges. It is a force much like the gravitational force between massive objects, though it can be both attractive and repulsive. When an electric charge moves, as when electrons move along a wire, an additional force is created. This is magnetism. The attractive force of a bar magnet is caused by an aligned and cumulative motion of the electrons in the iron atoms, each creating a tiny magnetic force and altogether creating enough force to pick up a paper clip. This kind of dynamic force that shows up only when the charged particles are in motion is perhaps the sort of thing that Mach had in mind for the effect of distant stars on the spinning water. He never worked out the details. That is, he never wrote down a precise law governing this putative effect. Magnetism, by contrast, is precisely understood and the laws of electricity and magnetism are well documented in Maxwell's equations. One of the consequences of Maxwell's equations is that an electric charge that is accelerated, wiggled up and down, say, will create an electro-magnetic wave. Moving a charge up and down will, because of the electric and magnetic forces of the moving charge, cause other charged particles at any distance to move up and down like a cork bobbing on a water wave. Electrons move up and down the antenna of a radio station and, by this electro-magnetic wave phenomenon, cause electrons in the antenna of your radio to move. These are radio waves. If the source electrons wiggle much faster than the radio source is capable, they can get the charged particles in the atoms of your skin to wiggle, an effect you feel as heat. Irradiated heat is also an electro-magnetic wave. So is light. Its frequency of oscillation is greater than that of heat, but it is still a case of wiggling charges at the source causing wiggling charges in your eye. That is how the energy is sent.

Just like a wave of water, it takes time for an electro-magnetic wave to travel from its source to the distant point where it has an effect. A charged particle that is further from the source will start to move later than a near particle, and it will always lag behind. Electro-magnetic waves travel much faster than water waves. No matter what the frequency, that is, whether it is a radio wave, radiant heat, or light, an electro-magnetic wave moves from source to reception at a speed of 3.0×10^8 meters per second. This is the speed of the wave in vacuum, that is, if there is no medium at all between source and reception. Through a medium like glass or air, the speed will be slightly less. Since light is the most popular electro-magnetic wave, people refer to this as the speed of light and give it the symbol "c". The speed of light, that is, the speed of all electro-magnetic waves, is thus

$$c = 3.0 \times 10^8 \text{ m/s.}$$

This value of the speed of light is actually part of the laws of electricity and magnetism. It appears explicitly in Maxwell's equations and you would not have to measure the speed of light to know it precisely. You can just read it in the laws of electricity and magnetism. This is incredibly important, and it marks the start of the special theory of relativity.

In what reference frame is the speed of light equal to 3.0×10^8 m/s? Doesn't it seem that the speed of light would be (3.0×10^8 m/s + 100 m/s) in a frame going 100 m/s toward the source of the light? Or in a frame going 3.0×10^8 m/s away from the source, catching up to and running with the light itself, the speed of the light would be zero? But if the speed of light is explicitly part of the laws of electricity and magnetism, then these laws would be relative, in violation of the principle of relativity because the speed of light, and hence the laws themselves, would be different in different reference frames. No other laws of physics include a particular *value* of a speed. Light, that is electro-magnetic waves, are peculiar in this way.

The logic of the situation is simple. The laws of electricity and magnetism cannot be absolute in the sense of being the same in all inertial reference frames unless the speed of light is absolute. In that case, said Einstein, the speed of light must be absolute. From this claim, and from the principle of relativity, all the rest of the special theory of relativity follows. These are the two foundations. Notice that both of these foundations of the special theory of relativity tell us about things that are *not* relative.

It is easy to say that the speed of light is the same in all inertial reference frames. It seems pretty innocuous, but it is really a stunning idea. If you click on a flashlight, the front of the beam moves away from you at 3.0×10^8 m/s. The front will also be moving away at 3.0×10^8 m/s from someone moving past you at just the moment you click on the flashlight, going 1.5×10^8 m/s, that is 1/2 c, in the direction you shine the light. Someone going practically as fast as light itself, say 2.9×10^8 m/s, will still have the front moving at 3.0×10^8 m/s away from them. There is no catching up to light. You cannot even get close, because no matter what your speed relative to the source, you will always measure the light going by at 3.0×10^8 m/s. That is what it means to say that c is absolute.

No other phenomenon is like an electro-magnetic wave, and no other speed is absolute. If you toss a ball with a speed of 3 m/s, then someone going by at 1 m/s in the same direction will measure the ball's speed to be 2 m/s. Someone going by at 3 m/s will measure it to be zero. They will have caught up to the ball. The speed of a ball is relative. All speeds are relative, except the speed of light.

The proof of a scientific claim is found both in its derivation and in its consequences. We can entertain the suggestion of an absolute speed of light and see where it leads, see what other phenomena it helps to explain. One of the earliest and most famous explanatory successes of the absolute speed of light hypothesis was in the Michelson-Morley experiment. If the speed of light were relative, as everyone believed before Einstein's suggestion to the contrary, then there must be a unique reference frame in which c has the value 3.0×10^8 m/s. Furthermore, since light is an electro-magnetic *wave*, and waves require a medium, something to move through, this unique reference frame must simply be the rest frame of the medium. The analogy to water waves is strong at this point. Water waves are just a disturbance in the medium, the water, and they move at some speed with respect to the stationary body of water. So electro-magnetic waves must be a disturbance in some medium and must move with speed c with respect to the medium. This hypothetical medium, which must pervade all parts of the universe through which light can travel, was dubbed the aether. Newton, if he had thought that light traveled as a wave, would probably have liked the idea of the aether. He would have said that the aether is at rest in space itself. Aether is the stuff of substantival space.

In 1887, Michelson and Morley tried to measure the speed of the earth moving through the aether. They did this by shining light in two perpendicular, horizontal directions. If one beam goes up-stream in the aether, and the other goes crosswise, they will take different amounts of time to cover equal distances. Michelson and Morley did the experiment repeatedly, at different times of day, different seasons, different locations, and turning the apparatus through a variety of orientations. There was never an appreciable difference in travel times of the two beams of light.

Einstein's suggestion of the absolute speed of light was not made to explain the Michelson-Morley null results. His motivation was the principle of relativity. But in fact Einstein's idea turns out to be just the thing to make sense of the Michelson-Morley experiment. There is no unique reference frame in which c is 3.0×10^8 m/s. It has that value in all frames. There is no aether. The speed of light is the same for both beams in the experiment, regardless of the motion of the earth.

CONSEQUENCES OF THE ABSOLUTE SPEED OF LIGHT

Einstein dramatically changed an aspect of physics: the description of the speed of light. Where the old, Newtonian physics described it as a relative property, the new, relativistic physics lists it as an absolute property. It should not come as a surprise that a change in one part of our model of nature will force some changes in other parts in order to maintain overall consistency. The Newtonian ideas of mechanics, that is, of motion and mass and forces, were predicated on the understanding that all inertial speeds were relative. We must now revise some aspects of mechanics in light of the fact that the speed of electro-magnetic waves is absolute. Some of the properties we used to think were relative will turn out to be absolute. Some we thought to be absolute will turn out to be relative. Rearranging this list is the challenge and revision to common sense that I warned you about earlier.

The key to understanding the special theory of relativity is the Minkowski diagram. All we have to do is add to the diagram the information that the speed of light is the same in all inertial reference frames, and see what happens.

Figure 3.5a is the Minkowski diagram for inertial reference frame K. Recall that the dashed line is the worldline of light and its speed is indicated by the angle it makes between the x-axis and the t-axis. It bisects the angle between x and t. Also note that the worldline of an observer (or an inanimate object, for that matter), stationary at the origin of K is exactly the t-axis itself. This is the line that shows no change in position at the origin.

Now we want to superimpose on this K-frame the Minkowski diagram for another inertial reference frame K' that is moving to the right from the K perspective. First draw in the K diagram the worldline of an observer at rest at the origin of K'. This would be, for example, someone sitting on the train, if K' is the train frame. This worldline must be at an

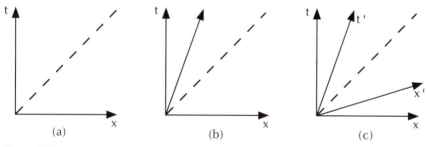

Figure 3.5

angle to the right since this person changes position according to K, as time goes on. Figure 3.5b shows the same K-frame diagram now with this worldline of someone sitting in the K′ frame added. Just as the worldline of the K-frame observer was the K time axis, the worldline of the K′-frame observer is the K′ time axis. Label it t′.

Do not forget that this is just a mathematical way of representing properties of time and position. It does not mean that real things on the train are slanted or listing. This is just how a consistent pursuit of the concepts has us represent things.

Still pursuing the concepts consistently, how should we draw in the x′-axis, the physical direction axis for K′? You might think it should point down and to the right, as if the t′- and x′-axes of the K′ system are just a rotated version of the K system's t- and x-axes. But this will not do because the worldline of the light ray emanating from the origin would not lie halfway between t′ and x′. In the rotated axis system, the light would be represented as going slower than it would be in the K system. The only way to have the speed of light the same in K′ and K is to draw the x′-axis rotated *up* from the x-axis, rotated by the same angle as t′ is rotated down from t. This way, the worldline of light bisects the angle between t′ and x′, as it bisects the angle between t and x. Any point on the worldline of light is as far along the x′-axis as it is along the t′-axis, so the speed of light in the K′ system is still 3.0×10^8 m/s.

Figure 3.5c shows the original K Minkowski diagram with the K′ Minkowski diagram added. This will be our tool for comparing quantities like speed, length, and duration of time in two different reference frames.

Again, note that the Minkowski diagram is very abstract. The fact that the t′- and x′-axes are squashed a bit does not mean that any physical object is squashed in the K′ system any more than writing the word "apple" in blue ink means we are talking about a blue piece of fruit. We have to draw the t′- and x′-axes squashed together to honor the fact that the speed of light is absolute, and this fact followed from a rigorous adherence to the principle of relativity. All of this information is now included on the Minkowski diagram.

The consequences of this information are made clear by showing which properties of things are relative and which are absolute. We start with an easy one in which common sense needs no help from physics. Consider two successive events A and B, like the bug's first bite of the eraser and his last, and ask whether the two events happen at the same place. More to the point, is the property of being at the same place relative or absolute? If one inertial reference frame measures the two events to be at the same place, will all other inertial reference frames agree? No. The property of at-the-same-place is relative, according to Newtonian physics, relativistic physics, and common sense. Here is an example at the level of common sense. I am so clumsy that I spilled coffee in my car twice in the same day. You may ask where, wondering if once it happened on the highway and then later in a parking lot or at a stoplight. At least the different circumstances would mean I did not make exactly the same mistake twice. But that's just it, I respond to your where-question. Both times were exactly at the same place, right into those little slots in the dashboard that empty into the radio. But where were you each time?, you clarify your question. Oh, the first was in my driveway, and the second was at a stoplight. In the car's reference frame, in other words, the two events were at the same place, the radio. But in a reference frame attached to the ground, the two events were not at the same place. One was on my driveway while the other was at a particular intersection. At-the-same-place is a relative property.

The Relativity of Simultaneity

Now consider two distinct events that occur at the same time. We have a word to describe this property: "simultaneous." The question is whether being simultaneous is relative or absolute. If, for example, events happen at opposite ends of the train car and observers on the train measure them to have occurred at the same time, will the platform frame measure them as simultaneous as well? The Newtonian model of nature says yes; simultaneity, unlike at-the-same-place, is absolute. The relativistic model says no; simultaneity is relative. This is a key difference between relativistic and non-relativistic physics, and finally the name "theory of *relativity*" is starting to make sense. The relativity of simultaneity is not a postulate of the special theory of relativity, nor is it a whim of Einstein's. It is a consequence of the absolute nature of the speed of light.

There are two ways to demonstrate that the relativity of simultaneity follows directly from the absolute speed of light, either by abstract mathematical deduction or by a physical experiment. Since physicists tend to do both, under the headings of theory and experiment, we will do both. Looking at the relativity of simultaneity in both ways will promote the understanding of the essence of the special theory of relativity, and this is what we need to know if we are dealing with appearance or reality.

The mathematical argument can be done entirely with Minkowski diagrams. These contain all the information of the complicated equations, the Lorentz transformations, that are usually used to compare properties between one inertial reference frame and another. We will do no calculations, and all we need is a ruler to get clear and definitive results.

We need to focus on two specific events A and B so we can draw these on the Minkowski diagram. Choose A and B to be events, firecrackers popping that happen at opposite ends of the train car, and that are judged to be simultaneous by an observer on the train. The train reference frame can be K, and we can sketch the particular details of this situation on a Minkowski diagram.

In the train frame the two worldlines of the two ends of the car are parallel to the t-axis since these two points are stationary with respect to an observer on the train. The instantaneous events A and B are points at a place and at a time. On any Minkowski diagram, all of the events that happen at time t = 0 are on the x-axis. All events that happen at time t = 1 second are on the line parallel to the x-axis that runs through the 1 second mark on the t-axis. Any events that happen at the same time are on the same line parallel to the x-axis. Some of the lines-of-simultaneity have been drawn onto the Minkowski diagram of Figure 3.6. Stipulating that A and B are simultaneous is stipulating that the line connecting them on the Minkowski diagram is parallel to the lines of simultaneity.

Now draw the Minkowski diagram of the platform reference frame K' that is moving with respect to the train, and use this diagram to evaluate the same two events. To avoid a confusion of lines, Figure 3.7 shows the K' diagram by itself, without K. Events A and B are there, unchanged, because it is the exact same physical events that we are talking about. The only difference is that the properties of events will now be measured with respect to the platform. It is still the case that the x'-axis represents all the events at time t' = 0. That is, the x'-axis is a line of simultaneity in K'. Other lines of simultaneity are parallel to the x'-axis, as lines of at-the-same-place are parallel to the t'-axis. K' graph paper would have a squashed appearance where the cross-hatching would be diamond shapes rather than squares. Sketching in some lines of simultaneity in K' shows that B lies on one line of simultaneity and A lies on another. A and B are not simultaneous in the K' reference system. B happens before A.

The mathematical consequences of the absolute speed of light are demonstrated by the Minkowski diagrams. In general, since the lines of simultaneity in one frame do not line up with those in another, events

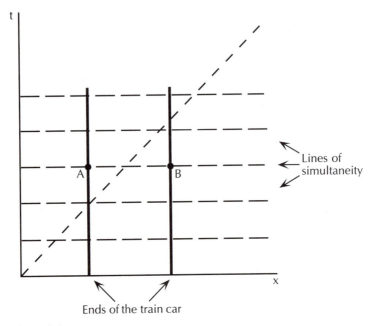

Figure 3.6

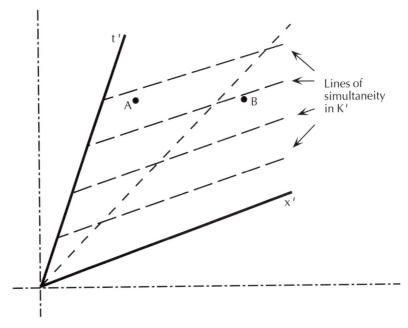

Figure 3.7

52

that happen at the same time in one frame happen in sequence in another. At-the-same-time is relative, just as at-the-same-place is relative.

It is worth doing the same analysis over again with more attention to the physical events themselves and the process of measuring simultaneity. Until it is clear *why* the events that are simultaneous in K are not simultaneous in K', the mathematics of a Minkowski diagram might not inspire all the confidence it deserves.

The distinction between measurement and perception is important, and it is measurement that we are more interested in. Perception is the brute sensation, simply what our sense organs report. Measurement is, all things considered, the reconstruction of events from the information given in perception and whatever we know already about the situation and about nature in general. We perceive a flash of lightening before the thunder. Nonetheless, knowing what we do about the speed of light and the speed of sound, careful measurement shows them to have occurred at the same time. We expect the most careful and considered measurement to be used as scientific observations.

The person on the train car who judges the events A and B to be simultaneous does so in full knowledge of the situation and the laws of physics. For example, an observer at the mid-point of the car who perceives the flashes simultaneously would say the events happened at the same time. An observer closer to one end of the car who sees the near flash just before the distant, could calculate, knowing the speed of light, to determine if the actual events happened at the same time.

The mid-car observer is the easiest to deal with. Figure 3.8 shows the physical events themselves at two moments of time. This is not an abstract representation of variables; this is a crude but realistic picture of what actually happens, as if we have taken two photographs. It is like a sequence of two photographs, the top one taken before the bottom. In the first snap-shot, the flashes of light are just leaving each of the two events A and B. The lower picture is of the event of the mid-car observer K seeing the two flashes. K sees the flashes at the same time, so the beams of light from A and B are arriving at the mid-point of the car at the same time. Figure 3.8 shows what happens for events A and B that are simultaneous in the K frame.

What does an observer on the platform make of all this? That is, how does an observer in another inertial frame K' that is moving to the right with respect to the train measure the simultaneity (or not) of the same two events? If K' is moving to the right with respect to K, then K is moving to the left with respect to K'. If you are standing on the platform, facing the track, the train is moving from right to left, and the event we call A is at the front of the car (the left end) and B is at the back. We know that the mid-car observer sees the flashes at the same time, or, eliminating the need for a human participant, the light flashes arrive at the mid-

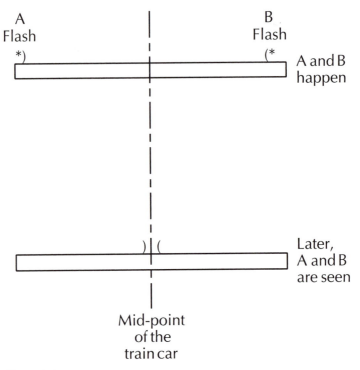

A
Flash

B
Flash

A and B
happen

Later,
A and B
are seen

Mid-point
of the
train car

Figure 3.8

point together. This is an objective, invariant fact, true in both K and K'. We can use a machine with a camera to record when each flash arrives at the mid-point, record it on a piece of paper and pass it to the person on the platform. No change of reference frame will alter the fact that the light from A and B arrived together at the mid-point of the car.

Knowing this, and knowing the laws of physics, K' must conclude that the events A and B were not simultaneous. Here is the reasoning: Since event A is ahead of the mid-car point and B is behind, the mid-point is moving toward the place where A happened and away from where B happened. In the time it takes for the light to travel from A to the mid-car K observer, that observer will have reduced the distance the light has to travel. The opposite happens for light from B. During the time the light is traveling, the distance it must go to reach mid-car is increasing. The key in this analysis, and the part that is distinctive to the special theory of relativity, is that, according to K' the speed of light does not get a boost from the speed of the source of the light. If it were tossed balls heading from A and B to the observer rather than light, then B's ball would get a boost and cover the extra distance with greater speed, and A's would have reduced speed and cover the shorter distance slower. They would arrive together. But this is light, and it travels back from A at exactly the same speed as it travels forward from B. Therefore, the only way that the two flashes could arrive at the mid-car point together, and we know that they do, is if the light from B got a head start. This is the only way that, going the same speed as light from A, the light from B could cover the extra distance and arrive on time. If A and B had happened at the same time then the mid-car observer on the train would have seen A before seeing B.

Figure 3.9 shows these events. These are exactly the same physical events reported from the train reference frame K and drawn in Figure 3.8. The experiment has only been done once. Firecrackers explode at opposite ends of a train car and the light from each arrives at mid-car at the same time. The two events are simultaneous in the train frame. They are not simultaneous in the platform frame. Simultaneity of events separated in space is a relative property.

It is not just that seeing events as simultaneous or not depends on your reference frame. That is, it is not just that the appearance of simultaneity is relative. This much is obvious and the special theory of relativity is saying much more. The point here is that the measurement of simultaneity, the reconstruction of what actually happened, is relative. Events *being* si-

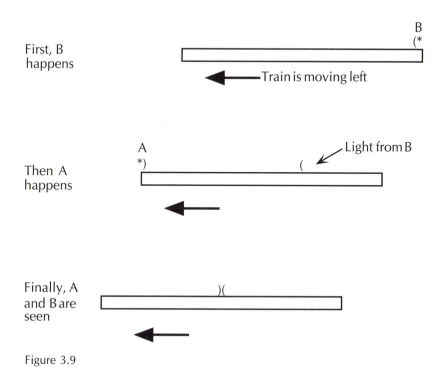

First, B
happens

Train is moving left

Then A
happens

A

Light from B

Finally, A
and B are
seen

Figure 3.9

multaneous, not just appearing to be simultaneous, is determinate only with respect to a particular reference frame. In this sense, a claim of the form "events A and B are simultaneous" is incomplete in the same way that "points a and b are on opposite sides of" is incomplete. On opposite sides of what? Simultaneous with respect to what reference frame? And just as the point of reference between a and b does not have to be a person, the frame of reference for A and B does not have to involve a person. It could be just another dot at point c that is between a and b. The train could be empty and the platform deserted. Nonetheless, if the flashes register together at mid-car, the events A and B are simultaneous in K, but in K', B happens before A.

The relativity of simultaneity reveals an interdependence between space and time that we failed to consider in the earlier suggestions about the nature of each. Simultaneity, that is, being at the same time, depends on motion through space. Change along one dimension forces a change along the other. Because of this interrelation, physics cannot consider one without the other. Relativistic physics cannot be about space or about time; it must be about spacetime. This does not mean that time is *the* fourth dimension. Time is not a physical, spatial dimension, and it is still very different from the three spatial dimensions. For example, we are free to move in any direction in space, left or right, up or down. But the dimension of time is, for some reason, more restrictive. Things can only go forward in time.

Time Dilation

Speaking of the passage of time, we can ask about another property of events, how much time passes between one happening and another. In language that is less substantival about time, is the property of time duration relative or absolute? In a casual, intuitive discussion of the bug eating the pencil eraser, the time that passed between the first bite (event A) and the last (event B) was thought to be an absolute property. Well, we were wrong. Newton was wrong too. As a consequence of the absolute speed of light, we are forced by the special theory of relativity to say that time duration is a relative property. The time that elapses in one inertial reference frame will be different from the time that elapses in another, between the same two events.

We have already proven a special case of this relativity of time duration by proving the relativity of simultaneity. In the K reference frame, events A and B are simultaneous, that is, the time duration Δt between them is zero. In the K' reference frame, some time does elapse between the two events, so $\Delta t'$ is not zero. In this sense, the time duration between A and B is relative.

This can be generalized to cases in which A and B are not simultaneous in either reference frame. The effect is called **Time Dilation**:

> The time duration between two events is longer in an inertial reference frame that is moving.

This is sometimes paraphrased by saying that moving clocks tick more slowly. In the rest frame of a clock, one second elapses between ticks of the clock. As reckoned in a reference frame moving past the clock, more than one second elapses between each tick of the clock. It ticks only once every second and a half, say, or once every minute, or every year. Its rate depends on the relative speed of the reference frame. And it is not just clocks that slow down. All physical processes exhibit the same time dilation. If my heart beats once every second, then to an observer in relative motion, it beats more slowly, maybe once every second and a half, or once every minute, or every year, depending on our relative motion. Since I will have a determinate, absolute number of heart beats in my lifetime, the moving observer will say that I live longer than what is measured in my own reference frame.

The amount of time dilation depends on the relative speed of the reference frame, and the difference in time duration is minuscule at speeds that we are used to. At 100 kilometers per hour, a thousand, or even a million kph, the effect is imperceptible without extremely precise clocks. It is only when the speeds approach that of light, 3.0×10^8 m/s, that time dilation becomes significant and noticeable. Some things in our world do reach such speeds, and with them the effect is observable. Tiny particles, the sub-atomic remnants of nuclear reactions in stars, stream into the earth's atmosphere from outer space. These are cosmic rays. Among them are particles called muons. While sitting in a laboratory, the half-life of muons is 2.2×10^{-6} seconds, but the identical muons flying by with a speed of 6/10 c have a half-life of 2.8×10^{-6} s. The muons actually exist for a longer time before decaying, according to a reference frame that is moving past them. In their own rest frame, according to the particles themselves, their half-life is always 2.2×10^{-6}. In the reference frame of the earth, a frame that is moving past them at 6/10 c, they live longer.

These are the physical manifestations of time dilation. We have yet to show how this relativistic effect is a direct consequence of the absolute speed of light. To do this will require a close look at how time duration is measured, that is, a look at the mechanism of a clock. To mark time, something in the clock must move, whether it is a swinging pendulum or oscillating atoms in a crystal. A very simple clock that exploits a known regularity in nature would involve shining a short pulse of light to reflect back and forth between two mirrors. Each time the pulse returns to the first mirror is a tick of the clock, and knowing the separation of the mirrors, and the speed of light, we can calibrate how much time elapses between ticks. If the mirrors are separated by the distance that light travels in half a second, that is, by 1.5×10^8 meters, then, since a round trip is twice this distance, each tick is a second. Figure 3.10a shows such a clock in its own reference frame K. Event A is the light striking the lower mirror. The pulse of light is reflected up to the upper mirror and reflected down again. The next strike on the lower mirror is event B, and the important property of this sequence is the elapsed time Δt between A and B.

In the rest frame K of the clock, events A and B happen at the same place. (They are separated by a little bit in the figure just to show the round trip and the two distinct events, A and B.) In a reference frame K' that is moving relative to the clock, A and B happen at different places. As a definitive example, suppose K' is moving horizontally past the clock, moving, that is, perpendicular to the path of the light pulse bouncing back and forth. The proof can be done for K' moving in any direction, but this choice makes things easier. If K' is moving to the left with respect to K, then from the K' perspective, the clock is moving right. Figure 3.10b shows that in the time it takes for the light to get from the bottom mirror (event A) to the top, the whole apparatus has moved some distance to the right. And in the time it takes to go from top to bottom, the mirrors have moved

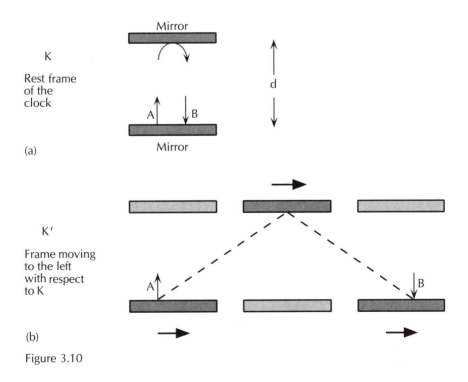

(a)

K

Rest frame of the clock

(b)

K'

Frame moving to the left with respect to K

Figure 3.10

some more to the right. In the moving clock, the light pulse has a longer path to follow, but the speed of light is the same whether the clock is moving or standing still. A longer path at the same speed takes more time. The time duration $\Delta t'$ between A and B is greater for the moving clock.

Notice that this effect is entirely symmetric. Let us say that K is the reference frame of the train, and there is a clock on the train. K' is the reference frame of the platform and there is an identical clock on the platform. An observer on the train will say that his clock ticks straight up and down but the clock on the platform ticks in a zig-zag. He will say the platform clock runs slow, taking more than a second between ticks. An observer on the platform will say *his* clock ticks straight up and down, and it is the clock on the train that ticks zig-zag. The *train* clock runs slow.

Length Contraction

The same sort of symmetry is true of length. Just as the temporal separation between two events is different in different reference frames, the spatial separation between two points will be different in different reference frames. The spatial phenomenon is called **Length Contraction**:

> The distance between two points will be shorter according to
> an inertial reference frame that is moving along the direction
> of the line between the two points.

An observer on the platform measures the length of something on the moving train to be shorter than an observer on the train measures. And an observer on the train measures things on the platform to be shorter than is measured by the observer on the platform. We can take two identical meter sticks, hold them side-by-side to check their equal length, and lay one on the platform along the tracks, the other on the train along the length of the car. When the train passes the platform, from the perspective of the platform the stick on the train will be shorter than a meter. From the perspective of the train car, the stick on the platform will be shorter than a meter. This is the symmetry; each reference frame sees the other's lengths as contracted. And again, this contraction is of noticeable magnitude only when the relative speed between reference frames is enormous, approaching the speed of light.

Newtonian physics and our initial, intuitive, analysis figured length to be an absolute property of things. But we might have anticipated that it would be a relative property in the special theory of relativity. It is not just the name of the theory that gives it away. In fact, that name has been rather deceiving on other issues. But the clue that length is relative is in the realization that the measurement of length involves either the measurement of simultaneity or of time duration, and both of these we already know to be relative. The length of a meter stick, or the train car, or whatever, is measured either by noting where the two end points are *at the same time*, or, if the object is moving, by noting *the duration of time* between the front end passing a certain marker and the back end passing the same marker. Since either of these measurements will vary from reference frame to reference frame, we expect the length itself to vary.

The length of the train car makes an easy example to show the relativity of length in general. The demonstration takes advantage of the work we have already done in understanding the relativity of simultaneity. A convenient way to measure the length of the train car as it is passing by would be to have the train pass through a tunnel of known length, and compare the length of the tunnel to the car.

Here is a specific setup. Send the train through a tunnel of length L. Event A is the instant that the rear end of the car is at the entrance (the left end) of the tunnel, and event B is the instant that the front end of the car is at the exit (the right end) of the tunnel. Let us choose a tunnel of just the right length such that events A and B are simultaneous in the rest frame of the tunnel. In other words, an observer at rest on the earth would observe the car enter the tunnel and at just the moment the tail end goes in, the front end comes out the other side. The car exactly fits into the tunnel. From this perspective, the train car is the same length as the tunnel. The car's length is L. Figure 3.11 shows this setup from the perspective of the tunnel's reference frame.

What is the length of the car in its own reference frame? That is, what would someone on the train measure the length of the car to be? You might want to consider the answer to this question yourself, just using the principle of length contraction, before we do the analysis. Length will be shorter from the perspective of a moving reference frame. The earth-tunnel frame is moving with respect to the train car, and so we expect the value of length in that frame will be less than the length in the car's own reference frame. We expect that the person on the car will measure the length of the car to be greater than the length of the tunnel, that is, greater than L.

This is exactly what follows as a consequence of the absolute speed of light. From earlier analysis, we know that events that are simultaneous in one frame will not be simultaneous in another reference frame. In the train reference frame, the events A and B are not simultaneous. In the train reference frame, the tunnel is moving by with the entrance ahead of the exit. The only way that the tunnel frame observer could measure A and B as simultaneous is if B happened *before* A. In the train reference frame, one end of the car exits the tunnel before the other end has entered. The car does not fit in the tunnel because the car is longer than the tunnel. The car is longer than L.

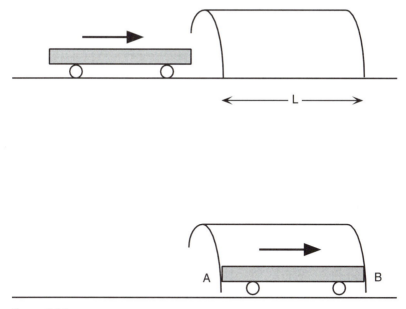

Figure 3.11

An observer on the ground measures the length of the moving car to be shorter than its length as measured by an observer on the car itself. The observer on the train car measures the tunnel to be shorter than its measured length by the observer on the ground. This symmetric length contraction is not a causal compressing of the objects. It is a different sort of phenomenon than compressing a sponge by squeezing it. The relativistic length contraction does not require a force of compression. The question of what force has compressed the train car does not even come up. It is exactly like the property of position. If, from my reference frame the pencil is sitting to the right, but from your reference frame the pencil is sitting to the left, we do not ask what force pushed the pencil from right to left. The relativity of length is the same. The car really is longer in its own reference frame, just as the pencil really is to the right of the origin of my reference frame. The change of length from one reference frame to another, just like the change of position or of speed, is in the real nature of measurement.

MITCH'S PARADOX

These relativistic effects are all direct consequences of the absolute speed of light. If the speed of light is absolute, then length, time duration, and simultaneity *must* be relative. This straightforward and plausible logical foundation not withstanding, the effects themselves may seem a test of plausibility. Length contraction and time dilation may seem to be a kind of descriptive slight of hand in which, simply by reorienting our perspective on nature, things *seem* to shrink and time *seems* to slow down. The symmetry of the effects in particular is unnerving. I say your time goes slow, and you say my time goes slow, and we are both right. This seems to invite the sort of contradiction that would show that we are not talking about the way things really are. At best we would be describing how things appear to different people.

The exposition of alleged or apparent contradictions in the special theory of relativity is usually done under the heading of a paradox. There is the famous Twin Paradox, the nearly as famous Pole-in-the-Barn Paradox, and others. These are all paradoxes in the sense that the situations they describe *seem* to entail contradictions, but on careful analysis there is no contradiction after all. They are all valuable cases to study because they challenge and sharpen the understanding of the special theory of relativity, and they show how its description of nature is entirely consistent.

One of the best paradoxes is not famous at all. It was suggested by my friend Mitch, so I call it Mitch's Paradox. It is inspired, as is the Twin Paradox, by the perplexing symmetry of time dilation. Mitch's Paradox involves two people born at the same time, but in significantly different circumstances than the Twin Paradox. The more famous case has the distinct disadvantage of involving a non-inertial reference frame and thereby going outside the domain of the special theory of relativity. Everything is inertial in Mitch's Paradox, and we can solve the initial puzzle quite easily.

There are two men, Bob and Bubba, who are exactly the same age and who live in neighboring towns along the railway line. They remain at rest in reference frame K, separated by a distance d (as measured in K). Another person, Richard, passes by, moving along the line separating Bob and Bubba. When Richard passes next to Bob, we note that Richard and Bob are exactly the same age. The challenging question is this: When Richard then passes by Bubba, will Richard be younger than Bubba, older than Bubba, or the same age as Bubba? The challenge is generated by the symmetry of time dilation. On the one hand, from the reference frame K of Bob and Bubba, Richard will age more slowly, so Richard will be younger than Bubba. But on the other hand, from Richard's own reference frame K', Bob and Bubba should age more slowly and Richard will be older than Bubba. Either one of these arguments is wrong (and we have to figure out which one), or the special theory of relativity contains a contradiction (and it cannot describe how nature really is). Richard cannot be both younger and older than Bubba. There is an absolute fact of the matter about their ages at the moment they are together. We can take a photograph of the two of them together as Richard speeds by, and either Richard or Bubba will have more gray hair and more wrinkles. This photograph will show unambiguously who is older, and it can be passed from one reference frame to another without alteration. The two frames must agree on who is younger at the event of Richard and Bubba being together.

The answer is that Richard is unambiguously younger than Bubba when the two are at the same point. Bubba will say that Richard looks younger than Bubba himself, and Richard will say that Bubba looks *older* than Richard himself.

This result does not violate the symmetry of time dilation because the procedures for measuring the time duration between the two events are not the same for the K frame and the K' frame. The two events in question are the meeting of Richard and Bob (event A) and the meeting of Richard and Bubba (event B). In Richard's reference frame K', the two events happen at the same place, namely at Richard's position. In the reference frame of Bob and Bubba, that is, in the K frame, the two events happen at different places, one at Bob's position and the other at Bubba's. The symmetry in the time dilation effect referred to a situation in which two different frames did *identical* time measurements on each other. Because of the difference in this case between the kinds of measurements done in the two frames, we should not expect symmetry.

The resolution of Mitch's Paradox can be explained either with a Minkowski diagram or a look at the physical situation. Figure 3.12 is a stylized picture of the physical situation, both in the K frame and the K' frame. The key to understanding why Richard is younger than Bubba when they meet is remembering length contraction. The distance between Bob and Bubba is d in the K frame, but it is shorter than d in the K' frame. As Richard moves by, the distance between Bob and Bubba contracts. So Bob and Bubba see Richard move at some speed for a length d. Richard sees Bob and Bubba moving at the same speed but the length between them is less than d. It takes less time for the Bob-Bubba segment to pass Richard than it takes Richard to move from Bob to Bubba. Thus, the duration of time between event A and event B is less for Richard than it is for Bob and Bubba. Richard ages less between these events. Richard is younger than Bubba.

But if everyone agrees that Richard and Bob are the same age at event A, and that Richard is younger than Bubba at event B, then, since Bob and Bubba are the same age, doesn't everyone agree that Richard is aging more slowly than Bob and Bubba? That is, isn't Richard's time absolutely slower? No, and the Minkowski diagram is useful for understanding why.

To say that Bob and Bubba are the same age is to say that they were born at the same time. This is an issue of simultaneity and so it is relative. The K' observer will not agree that Bob and Bubba were born at the

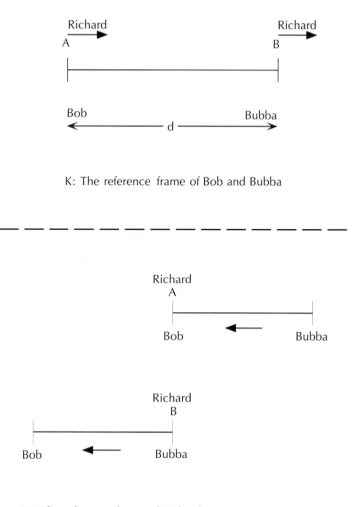

K: The reference frame of Bob and Bubba

K': The reference frame of Richard

Figure 3.12

same time since the events of their birth, though simultaneous in the K frame, are not simultaneous in the K' frame. As shown in Figure 3.13, using K' lines of simultaneity, Bubba was born before Bob, and Bubba is always older than Bob. That is why Bubba is older than Richard when those two get together. Bubba was older than Bob when Richard passed Bob.

The moral of this story is that the special theory of relativity does not lead us into any contradiction of the form that Richard is both younger and older than Bubba. When two people get together, however briefly, there will be a biological fact of the matter as to how much aging their bodies have done. Relativity does not ask us to ignore these biological, or other physical facts. If Bubba and Richard were piles of radioactive atoms instead of piles of organic molecules, we could note exactly how many in each pile had decayed at the moment of their being together. More of Bubba would have decayed, period.

NOTHING CAN GO FASTER THAN THE SPEED OF LIGHT

The speed of light is absolute, invariant in all inertial reference frames. The speed of light is also an upper limit on the speed of any causally effective physical signal. No object can go faster that the speed of light. No information in any form can go faster than the speed of light.

This second aspect of the speed of light, that it is a universal upper limit, is often cited as another postulate of the special theory of relativity. But we do not have to accept it as a postulate; it can be proven. Again, the Minkowski diagram is the key to clarity on this issue.

As a concrete example to work with, consider the speed of a baseball that is tossed toward a window. Say that event A is the event of the ball being tossed, and event B is the event of the ball hitting and breaking the

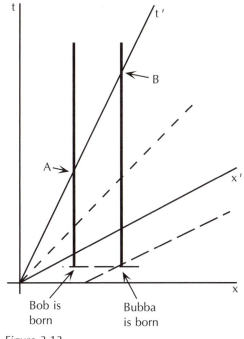

Figure 3.13

window. Clearly, event A must happen before event B, because A is the cause of B. The Minkowski diagram of this episode of vandalism is shown in Figure 3.14.

The speed of the ball is less than the speed of light, so the world line of the ball is at an angle less than 45° down from the vertical t-axis. If the speed of the ball was greater than the speed of light, its worldline would be more towards the horizontal than the dashed worldline of light. Notice on the Minkowski diagram that the time of the toss is indeed before the time of the window breaking. This will be true for any inertial reference frame. That is, for the two events in question, as long as the speed of the ball is less than the speed of light, all inertial reference frames will observe the same order of events, first A then B. Different frames will observe different amounts of time elapsed between the two events, but none will reverse the order.

If the speed of the ball were greater than the speed of light, there would be inertial frames in which the breaking of the window happens before the ball is tossed. In such a reference frame, the order of cause and effect will be reversed, and this is simply impossible. Thus, it is impossible for an object to go faster than light. This does not tell us the physical impediment to going faster than light. In this sense, the argument does not indicate *why* the speed of light is an upper limit. It is a logical argument of the form of a *reductio ad absurdum*. Speeds faster than light cannot be possible because that would lead to an impossible state of affairs, namely the cause of an event happening after the event itself.

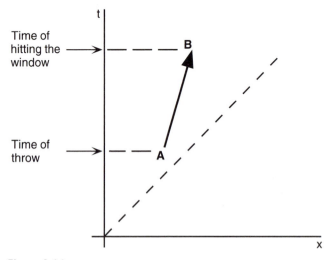

Figure 3.14

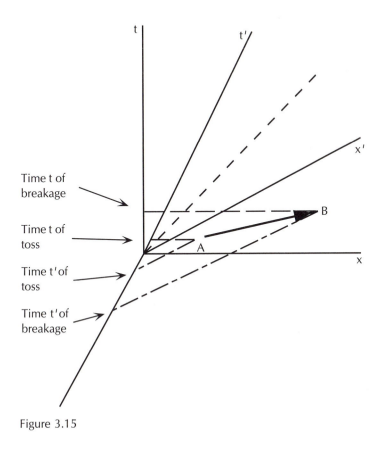

Figure 3.15

The Minkowski diagram in Figure 3.15 shows how faster-than-light worldlines can finish before they begin. In the reference frame K, the time of event A, the tossing of the ball, is indeed earlier than the time of event B, the breaking of the window. There is no problem with this. But in the reference frame K', the time of breakage is clearly before the time of the toss.

Nothing can go faster than the speed of light. Therefore, worldlines of real things must be steeper than 45°. Worldlines of real things cannot, at any point, be flatter than 45°. They can never be horizontal on the Minkowski diagram, since this would represent a change of position with no change of time, that is, an infinite speed. Worldlines of real things cannot loop around and cross themselves, since this would require some length of the worldline that dips below 45°.

The physical limit of the speed of light allows for a three-fold classification of the spacetime separation between any two events. Two events that can be connected by a straight worldline at a speed less than light are said to have a timelike separation. Events that are timelike separated can have a cause-and-effect relationship. The earlier can causally influence the later. On the other hand, if the straight line on a Minkowski diagram between two events is at an angle below 45°, then the separation between the two events is spacelike. There is no way that one of these events could affect the other, because any causal signal from one to the other would have to travel faster than light. The third possibility of separation between two events is that the straight line between them is at exactly 45°. This is a lightlike separation (sometimes called a null separation), and only a signal or object traveling at exactly the speed of light can link two lightlike-separated events together in a cause-and-effect relation.

Figure 3.16 shows examples of the three kinds of spacetime separation.

With this three-way classification in mind, the Minkowski diagram clearly shows the cause-and-effect possibilities of spacetime. If we put ourselves at the origin of the coordinate system, then the point where the time and space axes intersect is the instant here and now. All points below this are events in the past; all points above are events in the future. The lightlike worldlines, the 45° worldlines of light, separate the past into the spacetime region of those events that could have a causal influence on the here and now, and the region of those events that could not. A causal connection requires a timelike connection. Any event that lies outside the triangle of lightlike worldlines cannot be connected to the here

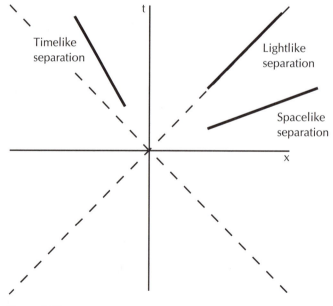

Figure 3.16

and now by a timelike signal, and so could not have a causal influence on us here and now. Not enough time has passed for a causal signal to cover the distance. Recall that our Minkowski diagram is simplified in that it does not show the other two spatial dimensions. On a fully generalized (three spatial plus one temporal dimensions) diagram, the triangle of lightlike lines becomes a three-dimensional cone. Any event outside the past light cone could not have effected the here and now.

There is also a future light cone formed by the lightlike worldlines that project into the future. Any event that lies inside the future light cone is an event we can influence. Any events outside the future light cone simply cannot be affected by anything we do here and now.

All of these restrictions, and the separation of spacetime into events that can and cannot be causally connected, are results of the upper limit on the speed of causal interactions. Nothing can go faster than the speed of light.

SUMMARY OF THE SPECIAL THEORY OF RELATIVITY

An effective way to summarize the important aspects of the special theory of relativity is to list the important properties of things in their proper category as relative or absolute. To clarify the distinction between relativistic physics and the common sense Newtonian physics it has replaced, those properties that have changed their relative/absolute status have been marked with an asterisk. So, for example, Newtonian physics describes length as an absolute property and the speed of light as relative. The special theory of relativity describes the speed of light as absolute and length as relative.

Summary of the Special Theory of Relativity

Relative Properties	Absolute Properties
position	laws of physics
speed	* speed of light
* length	
* time duration	
* simultaneity	

Even though the theory is called relativity, we need to pay attention to the column of absolute properties. They are, after all, the conceptual foundations of the theory. From a strict insistence that the laws of physics are absolute and the speed of light is absolute, all the rest of the special theory of relativity follows. The theory of relativity certainly does not say that everything is relative.

The entire discussion about relative and absolute properties has so far been restricted to comparisons between inertial reference frames. This is what is specialized in the special theory of relativity. Removing this restriction and generalizing the discussion to all kinds of reference frames brings us to the general theory of relativity.

Chapter 4
THE GENERAL THEORY
OF RELATIVITY

────────────────────────── • ──────────────────────────

The special theory of relativity depends on a distinction between inertial and non-inertial reference frames. A reference frame is inertial if it is not accelerating, that is, if it is at rest or moving at a constant speed, however fast, and in a straight line. We are forced to ask, especially now that we have done a little relativity, what is the frame of reference by which to judge this constant motion? Speed is a relative property, whether it is the speed of an object or the speed of a reference frame itself. In the context of relativistic physics then, describing a reference frame as inertial is an incomplete description unless we know the fundamental reference. We need to know inertial with respect to what?

Special relativity seems to require that one reference frame in the universe, or at least one class of reference frame, be singled out as the standard of all other inertial frames. This would put us on the road to a substantival model of space. It is exactly the result Newton claimed from his spinning bucket experiment, and it is exactly contrary to the relational account of space as advocated by Mach. The unique reference frame for distinguishing inertial from non-inertial motion must be the reference frame of space itself. An inertial frame is at rest or in uniform motion with respect to space itself.

As a historical note, Einstein was strongly motivated by Mach and the relational model of space. The idea behind relativity was to avoid any reference to a substantival space and to write physics entirely in terms of things and properties more directly observable. Since the only motion we can observe is motion relative to other objects, that is the only sort of motion physics should be concerned with. One way to avoid the reference to motion with regard to space itself would be to eliminate the need for a distinction between inertial and non-inertial reference frames. This is the project of general relativity. The idea is not to undo or negate the results of the special theory of relativity, but to expand them beyond the narrow setting of inertial reference frames. It will still be the case that the speed of light is absolute, but now the hope is to develop a theory in which absolute means invariant in *all* reference frames, not just inertial frames.

GENERAL COVARIANCE AND THE PRINCIPLE OF EQUIVALENCE

The principle of relativity is the philosophical heart of the special theory of relativity. When applied to the laws of electricity and magnetism, the principle forced a change in our understanding of the speed of light, and from then on all the other aspects of the theory were logical consequences. The generalization of the principle plays the same fundamental role for the general theory of relativity. The generalized claim is called the **Principle of General Covariance**. It is simply the principle of relativity with the word "inertial" erased:

The laws of physics are the same in all reference frames.

The name general covariance is less misleading than Principle of Relativity. The meaning of "covariance" is similar to invariance. Covariance refers to the form of a law, whereas invariance refers to the value of a specific property like speed or mass or length. So a law is generally covariant if its form is the same in all reference frames. You do not have to add any additional forces or sources of energy in one reference frame to balance the equations formulated in another. The equations are the same, though specific values of specific variables may change. This is covariance.

We do not have to look very far to find an apparent counterexample to the principle of general covariance, no further, in fact, than the bagel in the lap of the napping businessman on last chapter's train. The law of physics we are concerned about is that objects accelerate only when under the influence of an external force. No force, no acceleration. $F = ma$. This law works flawlessly on the train as long as the train itself is not accelerating. A ball put at rest on a level table on the train will remain at rest on the table as long as the train goes at a constant speed. No force, no acceleration. But when the engineer puts on the brakes and the train begins to slow down, the ball spontaneously begins to roll across the table. No one has pushed the ball. There is no electric or magnetic force in effect. As the ball begins to move on its own and picks up speed across the table, this seems to be a case of acceleration with no force. If this is true, then the law $F = ma$ is not true in the non-inertial frame. $F = ma$ does not seem to be generally covariant.

We should not be discouraged by this. Recall that in developing the special theory of relativity it was right after introducing the principle of relativity that a challenging counterexample came up. The laws of electricity and magnetism seemed not to be the same in all inertial reference frames because of the speed of light. The solution to this problem was the start of great things. It was just the challenge to launch the new theory. Pursuing a similar approach then, Einstein asked what modification can be made to our old way of thinking about nature that will bring the law of acceleration and force into compliance with the principle of general covariance.

Thinking back to the ball spontaneously rolling across the table on the train, a similar thing happens to us all the time in the inertial reference frame of our room. In a room where F = ma certainly *is* true, if you are holding the ball in your hand and suddenly let go, the ball spontaneously falls to the floor. There are no springs connecting the ball to the floor. No one is pushing or pulling the ball. There is no wind blowing it down. No visible force is acting on the ball, and yet it accelerates. When this apparently spontaneous acceleration happened on the train it was because the train itself was accelerating. The ball moved forward when the train slowed down. So in the case of the ball accelerating down toward the floor of the room, is it because the room itself is accelerating up? No. As the cause of the ball's acceleration we cite gravity, a force for which we have no evidence other than these sorts of accelerations. But with the force of gravity in the equation, F = ma is an accurate description of the situation. The force of gravity is itself caused by the mass of the earth, and so what seemed at first to be a non-inertial phenomenon—the spontaneous acceleration towards the ground—is explained by reference to an object, not to space itself.

The goal of relativity is to generalize on this case so that all inertial/non-inertial distinctions in all cases are determined by the objects in space with no reference to space itself. Toward that generalization, recall that in the case of the falling ball, things would have happened exactly the same if there were no gravity but the room, the reference frame, was accelerating upwards. It is easier to imagine this happening in an elevator than in a room. In outer space, with no massive planets or stars nearby to create a gravitational force, if the elevator is accelerated by its cable or by having rockets firing from below, things will accelerate to the floor, just as things fly to the back of an airplane as it picks up speed down the runway. In other words, the effects of gravity can be perfectly mimicked by the acceleration of the reference frame, at least in this small, localized region of the elevator. So what we call an accelerating frame, that is, a non-inertial frame, can be characterized in terms of gravity and hence by reference to other objects and not to space. This idea is the second pillar of the general theory of relativity. It is called the **Principle of Equivalence**:

> The effects produced by an accelerating reference frame are indistinguishable from the effects produced by gravity.

This fits nicely with the fact that all objects, regardless of their mass or their composition, fall at the same acceleration under the influence of the earth's gravity. The rate of acceleration is roughly 9.8 meters per second, per second. That is, a dropped object picks up speed, adding 9.8 m/s to its speed in each second it is falling. Since the effect of gravity is indistinguishable from the effect of an accelerating reference frame, dropping a brick and a penny on the earth must be like letting go of a brick and a penny in an elevator going up, picking up speed at a rate of 9.8 m/s per second. In the elevator case, the brick and penny just sit there while the floor rushes up to hit them, so of course they hit the floor at the same time.

Just like the principle of the absolute speed of light, the principle of equivalence has been introduced as a way to maintain the covariance of the laws of physics in the face of seemingly recalcitrant examples. If we can cite gravity as the cause of non-inertial effects, and use the gravitational objects in the universe as the reference for distinguishing inertial from non-inertial reference frames, we eliminate the need to cite substantival space and we find real causes, real forces for all observable acceleration. Also, just like the principle of the absolute speed of light, the principle of equivalence leads to some challenging consequences. The phenomena predicted by the principle of equivalence are what are most commonly identified with the general theory of relativity. The predictions are also the way to test the basic foundations of the theory.

CONSEQUENCES OF GENERAL COVARIANCE AND THE PRINCIPLE OF EQUIVALENCE
The Bending of Light

In a stationary reference frame, light always travels in a straight line. This is true as long as the medium is uniform, that is, the light does not pass from air to water, or air to glass, or some other this to that. If the medium changes like this we have to worry about refraction effects as in a lens. So for all of the experiments we do, assume the medium is uniform. For simplicity, assume it is going through air, or even through a vacuum.

In a stationary reference frame, light always travels in a straight line. Put yourself in a dark box with a small hole in one wall near the top. A short pulse of light from a flashlight, projected horizontally in through the hole will hit the opposite wall at the same level as the hole. The path of the light is a horizontal straight line. If your box is moving up at a constant speed, an inertial box, then, in the time it takes for the light to cross the box, the opposite wall will have moved up some distance and the light will hit the wall at some point below the point across from the hole. Even in this case though, the path of the light is a straight line. It is just at an angle, down from the hole to where it hits the opposite wall. If you note the intermediate points taken by the light, you will see they fall on a straight line.

Now do the same experiment a third time but have the box accelerating upward. In practice this could be done by fixing the box up like a rocket with the engines on the bottom. When we say that the box is accelerating we mean that the engines are on and are giving the box a steady push. The box is constantly picking up speed. In the time it takes the light to go halfway across the box, the box has gone up some distance. But in the remaining time to get to the opposite wall, the box goes a greater distance because, picking up speed, it is now going faster. Figure 4.1 shows this sequence of events, both from the perspective outside the box and from inside the box. Inside the box, that is, in the accelerating reference frame, you see the light hit the opposite wall at a point below what is opposite the hole. Furthermore, if you note the intermediate points taken by the light between the hole and the wall, it is a curved path. The light does not angle down; it bends down. In an accelerating frame, light rays bend.

In a substantival model of space, this is pretty easy to explain. Light itself, you could say, always goes in a straight line. That is the law of

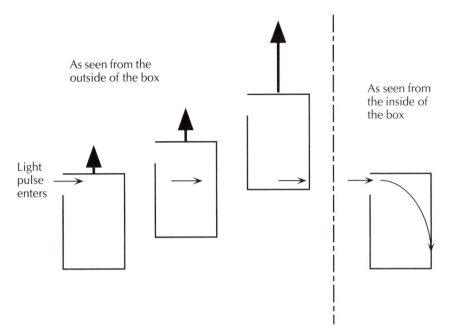

Figure 4.1

physics. In the accelerating box, the path of the light only *seems* to be curved. It is our accelerated way of looking at it that gives the light this appearance. In reality, the trajectory is straight. And the moral of the story is that the accelerated reference frame, accelerated with respect to space itself, is poorly suited for studying physics. Non-inertial frames are susceptible to just these sorts of spurious effects.

With the principle of equivalence though, the bending of the light is no more illusory, no less real, than the acceleration of the ball that falls to the floor in a room on the earth. In fact, the principle says that the bending ray of light is, for all observational purposes, the same phenomenon as the falling ball. If the effects of an accelerating frame are indistinguishable from the effects of gravity, then light passing into and across a box sitting still near a very massive object will also bend. It will bend down toward the gravitational source, just as a tossed ball arcs down to the ground. Figure 4.2 shows the observational equivalence of the two situations, the accelerating box and the box with gravity.

In either case, the light travels very quickly from one side of the box to the other. For this reason, it takes an incredibly high rate of acceleration,

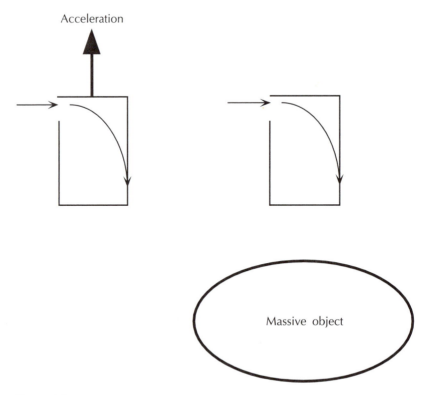

Figure 4.2

or an incredibly strong force of gravity, to make the light ray bend appreciably. The 9.8 m/s per second acceleration of gravity here on the earth does not cause much bending at all, certainly not enough for us to see.

The precision with which Einstein's general theory of relativity predicts the bending of light by gravity make this phenomenon a natural place to test the theory. Such a test would require a hugely massive object with a gravitational force much larger than the earth's. The natural candidate is the sun, the largest, most massive thing around. But the sun has a drawback in this experiment, and that is its own brightness. We do not want to measure the trajectory of light that comes from the sun itself, but of light that comes from somewhere else and passes close to the sun. Light will not bend when it shines straight up from the gravitational source any more than it will bend when it shines directly in the direction of the acceleration of the box, that is, straight up. It only bends when it shines crosswise, that is, when it passes by an object like the sun. But the sun's own light will outshine any light from a distant source that passes near the sun, except in the event of a total solar eclipse. Under eclipse conditions, light from the sun itself is blocked by the moon, while light from a distant star can pass close to the sun and arrive at an observatory on the earth. These are the circumstances to test the general theory of relativity.

The first feasible opportunity of such a test came in 1919. By noting the apparent positions of familiar stars as they appeared during an eclipse, astronomers could tell that the light rays from these stars did indeed bend as they passed by the sun, and by just the amount predicted by the general theory of relativity. The sun's gravity not only pulls the planets into orbit, it also pulls the passing light into a curved trajectory. Perhaps, with a gravitational source much stronger even than the sun, light would be bent so much as to loop around and orbit as the planets orbit the sun.

The phenomenon of light bending under the influence of gravity is a consequence of the principle of equivalence and the principle has been suggestive of a way to condition the inertial/non-inertial distinction on the objects in space rather than on space itself. The gravitational bending of light thus counts as evidence of a genuinely relativistic theory of motion and forces, and of a relational model of space. But, in order to come to a well-informed conclusion as to the nature of space and time in the general theory of relativity, we need to look at more evidence, and evidence of different kinds for the principle of equivalence.

Gravitational Red-shift (Time Dilation)

Gravity has the effect of bending rays of light. This is proven by first showing that an accelerating frame of reference has this effect and then invoking the principle of equivalence. Using the same strategy we can show that gravity has the effect of slowing down time. Clocks, and all other physical processes, run slower when under the influence of a strong gravitational force than when they are not.

It is worth noting right from the start of this discussion that this gravitational time dilation is significantly different from the time dilation of the special theory of relativity. The gravitational effect is not symmetric in the way that time dilation between inertial reference frames is. Between two inertial reference frames, each measures the other's time as going more slowly than their own. In the general relativistic effect, on the other hand, an observer far from any source of gravity will measure the time of someone near the source as going more slowly than his own, and the observer near the source will measure the distant time as going *faster*. This asymmetry will be shown in the proof of the effect, just as the symmetry in the special relativistic case is revealed in its proof.

Get back in your box, but close up the opening through which the flashlight was shining in the previous experiment. For this experiment we need two identical clocks, one at the bottom of the box, the other at the top. We are going to ask about the rate at which the bottom clock is ticking as observed by someone there with it at the bottom of the box, and as observed by someone at the top of the box. To facilitate the observation at the top, have the bottom clock send a beam of light to the top, and have the frequency of the light correlated to the frequency of ticks of the bottom clock. Recall that the frequency of light is a feature of the wave. The number of wave crests that pass a point of observation in a second is the frequency of the light. The idea of this experiment is for the observer at the top to be able to measure the rate of the clock at the bottom by measuring the correlated frequency of the light.

For reasons of efficiency in the discussion of the rates of clocks and the frequency of light, use the following abbreviations:

f_b = the frequency of the bottom light source as measured by the bottom clock

f_t = the frequency of the bottom light source as measured by the top clock

Note that both variables of interest are about the source of light at the *bottom* of the box. The subscripts indicate the different reference points for measuring this same thing.

First do the experiment with the box moving up at constant speed, that is, not accelerating. This can be done by firing the rockets for a while to get the box up to speed and then turning them off to let the box drift.

With the rockets turned off, the box moves at a constant speed. In this case, the clock at the top is always moving at the same speed as the clock at the bottom. There is no relative motion between top clock and bottom clock, and so the frequency of the light is the same when it is received at the top as it was when it left the bottom. That is, f_t and f_b are equal. The clock at the bottom is ticking at the same rate as the one at the top.

Now do the experiment with the box accelerating up, that is, constantly picking up speed. We should focus our attention on two events: the moment the light signal is sent from the bottom clock and the later moment when the signal is received and measured at the top. The light is sent with frequency f_b, as measured by the bottom clock, and received with frequency f_t, as measured by the top clock. Now, since the box is always picking up speed, the clock at the top will be moving faster when it receives the light than the clock at the bottom *was* going when it sent the light. In other words, in the accelerating box shown in Figure 4.3, the reception of the light is moving relative to the source of the light. This is different from the non-accelerating, constant-speed box.

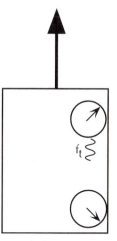

Later, light is received
at the top with frequency f_t

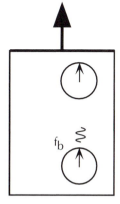

Light is sent from the
bottom with frequency f_b

Figure 4.3

Moving away from the source of light does not change the speed of light but it does change the wavelength and the frequency. This change is called the Doppler effect. The analogy to water waves is imperfect in this case, since the speed of a wave on water is relative, but we can use it to get the point across. Sitting on the beach with your feet just in the water, you feel the waves hit with a certain frequency, maybe one wave per ten seconds. If you leap up and swim out into the water, that is, move toward the source of the waves, the frequency of waves increases. You encounter more waves per second. Now turn around and swim back to shore. Moving away from the source of the waves, the frequency decreases. You encounter fewer waves per second.

The clock at the top of the accelerating box is moving away from the source of light. Therefore, the frequency of the light will be less as measured at the top than as measured at the bottom. f_t is less than f_b. Since the frequency of the light is correlated to the rate of the *bottom* clock, the observer at the top, who measures the bottom clock as producing a reduced-frequency signal, will say that the bottom clock is running slow. The greater the acceleration of the box, the greater the Doppler shift of frequency, the slower the bottom clock. And if the top clock in the accelerating box sent down a light signal with a frequency correlated to its rate of ticking, the bottom clock would be moving *toward* the source of light and hence would encounter a higher frequency than was sent. The observer on the bottom of the box will measure the clock at the top to be running fast.

If this phenomenon happens in the accelerating reference frame of the box, then, by the principle of equivalence, it also happens in the presence of a gravitational source. Accelerating up is like having a massive source of gravity beneath you. Once again, the magnitude of the effect—the amount of difference in the rates of the two clocks—depends on the magnitude of acceleration or the strength of gravity. It is not the sort of thing we could observe with regular clocks from the store, putting one on the floor of your room and one on a shelf. Nonetheless, even in the mild gravitational field of the earth, equivalent to an accelerating frame at 9.8 m/s per second, the clock that is closer to the source of gravity, the clock on the floor, goes a little bit slower. This gravitational time dilation has in fact been measured on the earth. Using identical and precise clocks separated by a few hundred feet from the top to the bottom of a tall building, the clock at the bottom goes more slowly, as predicted.

It is, of course, not just clocks that go slow near a source of gravity. The frequency of the light sent from the bottom of the box, whether the box is accelerating or near a source of gravity, could be correlated to any physical process, a clock, a heartbeat, a bug eating an eraser, whatever. Things in general happen more slowly at the bottom of the box than at the top. People who live at sea level age more slowly than people living in the mountains. It is not worth moving down to the shore though, and not only because the difference is too small to measure. There is no point in moving because, even if the effect were substantial you could not notice. It is a comparative effect. Time goes slower at the lower elevation *compared* to time at the higher elevation. Wherever you are, all physical phenomena will be progressing at the same rate. There will be no local reference by which you are aging more slowly, and you will not perceive any greater longevity. If you had not read about it here, you would never know.

The effect of time dilation near a gravitational source is called Gravitational Red-shift. A word of clarification is in order. The different colors of light that we see correspond to different frequencies of the electro-magnetic wave. The lowest frequency we can see is the color red. Lower frequency electro-magnetic radiation we cannot see but can feel as heat, hence the term infrared radiation in describing heat. The highest frequency we can see is violet. Between these two extremes are all the other colors in the order we see them in a rainbow. If visible light is sent from the bottom clock, its frequency will be reduced when observed from the top. Its color will be shifted toward red. This does not mean it will always *be* red. It could be violet at the source and shifted to blue at the top. Red-shift just means a shift to a lower frequency. It could even start out red and be shifted into the infrared. This would still be a red-shift. The gravitational red-shift then is the phenomenon of physical processes going slower in a strong gravitational field as compared to processes in a weak field. Slower physical processes mean slower oscillations of the charged particles that cause an electro-magnetic wave. The frequency of the wave is reduced. The wave is red-shifted.

The Curvature of Spacetime

The curvature of spacetime is a tricky subject and it would be best to point out some of the important questions to keep in mind as the issues unfold. First of all, there is the question of relational and substantival space and time. If we talk about the curvature of space and time, and we will, does that not presuppose that there is something there to be curved? In other words, does the concept of curved spacetime force a substantival model of space and time? A curved road has to be a real, substantival road. In the same way, perhaps, a curved spacetime must be more than a set of relations among physical objects. Please note that by asking this question at this point I do not mean to answer it now. The idea is to have something to think about as we go through the details of what it means to say that spacetime is curved. A second issue is put here for the same reason.

The second issue, related to the first, is that of appearance and reality. Is it more accurate to say that the results of measurements of things moving through space and time are *as if* the spacetime is curved, or that these measurements show that the spacetime *is* curved? Is the curvature a feature of the process of measuring or of the thing measured? Is it merely the appearance of curvature, or is the curvature a reality?

With these questions in mind, we can start the analysis of curvature with a heuristic example that Einstein himself used to make the transition from special to general relativity. It is a thought experiment in the spirit of the special theory of relativity, with two reference frames K and K'. They are not inertial reference frames. Instead of a Minkowski diagram, a space-space diagram is useful for showing the relevant features of the situation. Have the origins of the two coordinate systems coincide, and have the two systems in constant rotation with respect to each other. This rotation is what makes things non-inertial. Let us say that K' rotates counter-clockwise with respect to K. Equivalently, K rotates clockwise with respect to K'. Figure 4.4 shows a snapshot of the situation at one instant.

A real, physical construction of this setup would be easy. On a merry-go-round, draw perpendicular horizontal axes through the center and label them x' and y'. This is the K' coordinate system. Also draw perpendicular horizontal axes on the ground. They too intersect at the center of the merry-go-round. Label these x and y, the spatial axes of the K reference system.

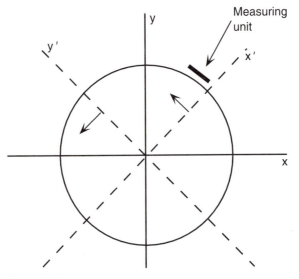

Figure 4.4

Now consider a circle that has its center at the origin of the two coordinate systems. This does not have to be any physical object. No material thing has to be shaped into a circle and placed just so. We are only interested in the set of points that are all the same distance from the center. This is a circle. The experiment entails measuring the diameter and the circumference of the circle. Do this first in the K reference frame by using a unit measuring stick. Lay it end to end along the diameter to measure d and along the circumference to measure c. The results will be related in the basic formula of Euclidean geometry, c = πd. The number π, in other words, is the ratio between the circumference and the diameter of a circle.

But what measurements does K' get for the diameter and circumference of the same circle? From the perspective of K, the unit measuring stick does not contract while it is laid along the diameter, but it does contract when laid along the circumference. This is because the stick is moving along the direction of its length when measuring c', but moving perpendicular to its length when measuring d'. Length contraction happens only in the direction of motion. So the two reference frames get the same measurement of diameter. d = d'. But the length-contracted stick will have to be laid down more times to get around the circumference in K', so the measurement of circumference in K' will be *greater* than in K. c' > c.

What is really crucial here is that in the accelerating K' reference frame, the formula of Euclidean geometry is no longer true. In the K' frame, c' > πd'.

This does not mean that the Euclidean formula for the relation between a circle's circumference and its diameter is false. It simply means it is inappropriate for describing the geometry of an accelerating reference frame. And there is nothing really wild or hair-raising about this. We encounter situations on a daily basis in which the Euclidean formula is inappropriate. In fact, we live on one, our spherical planet. Consider drawing a circle on the surface of a sphere. To focus your imagination, use the North Pole as the center of a circle drawn on the surface of the earth. If the circle is small compared to the size of the sphere, then measurements of the diameter and circumference will be nearly in a ratio of π. But as the circle gets bigger, the Euclidean formula fails. Drawing circles around the North Pole amounts to drawing lines of latitude, and bigger circles get you closer and closer to the equator. Think of the circle with a radius equal to the distance from the North Pole to the equator. This is shown in Figure 4.5. The circumference is the equator itself.

Since the radius of the circle goes one-fourth of the way around the earth, the diameter, which is twice the radius, is one-half of the way around the earth. The circumference, the equator, is exactly once around the earth. Thus, the diameter of this circle, as you would measure it on

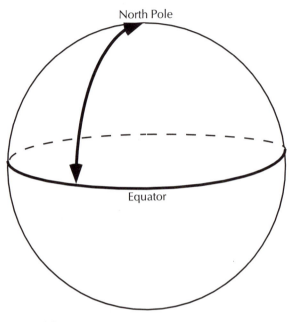

North Pole

Equator

Figure 4.5

the curved surface of the earth, would equal half the circumference. c = 2d, *not* c = πd. The Euclidean formula is inappropriate for describing geometric relations on the curved surface of the earth. Euclidean formulas are applicable only for geometric figures on flat surfaces like a desk top or a blackboard. For small figures on the earth, like the relatively small circles we started with, the Euclidean formulas are an adequate approximation. It is only on the large, global scale that the curvature of the earth is a notable factor and Euclidean geometry becomes the wrong way to describe things. The earth is locally flat, but globally curved.

If you were wondering about the earth, whether it was flat or spherical, you could find out by doing geometry. Draw a large circle and measure the diameter and the circumference. If the Euclidean formula c = πd is true of your measurements, things are flat. If your measurements do not fit the Euclidean formula, that is if the ratio of the circumference and the diameter is anything other than π, then the conclusion is that the earth is not flat. The surface on which the measurements were done is a curved surface.

Measurements of diameter and circumference in the accelerated reference frame K′ in Einstein's example did not fit the Euclidean formula. Geometry in the accelerating frame was non-Euclidean. If we follow the analogy with geometry on the earth to its conclusion, the non-Euclidean nature of measurements in the accelerating frame indicates that the surface on which the measurements are done is not flat. The space is curved.

The principle of equivalence says that if acceleration has the effect of a curvature of spacetime, then so does gravity. If, instead of rotating the reference frame K' there was a substantial gravitational source at its origin, measurements such as the one of diameter and circumference will be non-Euclidean. Massive objects, in other words, curve the spacetime around them, and the evidence for the curvature is in geometrical measurements. This prediction needs to be tested, just as the bending of light and the gravitational red shift were tested.

To facilitate such a test, and to further our understanding of the curvature of spacetime, we should return to earth and clarify a few more ways to measure the curvature of the planet. On a flat surface, the sum of the interior angles of any triangle will be 180°. This is a formula of Euclidean geometry. A triangle is, of course, a figure formed by three *straight* lines. If you lay out a large enough triangle on the surface of the earth though, measure the three interior angles and add these three values together, the sum will not equal 180°. To give an example, we must first introduce an unambiguous concept of a straight line on a curved surface. The concept should apply as well to what we already know to be a straight line on a flat surface. We clearly have an intuitive notion of a straight line on the earth. If you go from New York to Boston via Chicago, your line from New York to Boston has not been straight. The straight line is "as a crow flies," no deviation to one side or the other. The straight line is the shortest distance between the two points, and this is the definition we can use. Since "straight" seems to imply a line on a flat surface, we generalize the concept of the path of shortest distance under the term "geodesic." A geodesic is a line of shortest distance. Between New York and Boston there is a unique line that is the geodesic, just as, between any two points on a flat surface there is one and only one straight line. The geodesics on a sphere lie on the great circles of the sphere. So if you went from New York to Boston on the geodesic and you kept going straight, you would go all around the globe and end up back in New York.

A triangle on the earth must be constructed with the curved-surface equivalent of a straight line, namely with geodesics. Imagine the following triangle: One corner of the triangle is on the equator, and another is the North Pole. The straight line between these points, the shortest distance, the geodesic, runs straight up from the equator to the pole. This line is perpendicular to the equator. At the North Pole, rotate by 90° and drop another straight line back down to another point on the equator. The triangle is completed by connecting the two equatorial points with the unique geodesic between them. This lies on the equator. Figure 4.6 shows this triangle.

Now add the interior angles. Both lines from the North Pole to the equator are perpendicular to the equator, so each base angle is 90°. Add those two together and that is 180° already. Whatever the size of the angle at the pole, adding it in will push the sum over 180°. We chose 90° at the top just to be definitive. So the sum of the interior angles of the triangle drawn in Figure 4.6 is 270°, not 180°. This is another indication that the surface, the space, in which the geodesics are drawn and the measurements are made is non-Euclidean. It is a curved, two-dimensional space.

The concept of a triangle depends on a definitive notion of a straight line. So does the idea of parallel lines. On a flat surface, that is, in Euclidean geometry, if there is a straight line and a point somewhere that is not on the line, then there is exactly one other straight line that passes through that point and is parallel to the original line. The two lines, extended for unlimited length, will never intersect. But try the same construction on a sphere like the surface of the earth. Draw a geodesic and a point somewhere not on the line. There is no parallel geodesic that can be drawn through the point. There are no parallel lines on a sphere, where parallel means straight lines, geodesics, that never intersect. What you might think are parallel lines on the globe, lines of latitude like the tropic of Cancer and the tropic of Capricorn, are not geodesics. They are not straight. They do not intersect because they bend away from each other. The straight lines, the geodesics, the great circles, all intersect like longitudinal lines at the poles.

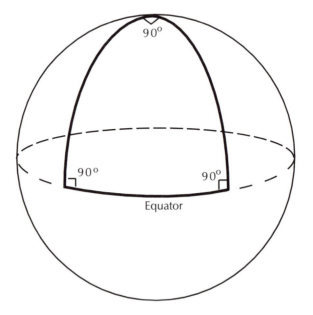

Figure 4.6

In this sense, a geodesic bends with a curved space, but it is the straightest possible line between two points in the curved space. The curvature of the geodesic can be observed and measured by observers and processes that are confined to the space itself. That is, you do not need an external vantage point to observe the bending. It is not just bending with respect to some other space or some other reference frame. The curvature and the bending are intrinsic to the earth itself. The curvature of the space and the bending of geodesics that goes along with it are absolute properties.

So here is a way to test the curvature of spacetime as predicted by the principle of equivalence. We can look for some intrinsic effect of curvature, some measurable behavior of geodesics, that takes place in the four-dimensional spacetime in which we and all of the other furniture of the universe are situated. The process is analogous to measuring circles or triangles on the two-dimensional surface of the earth to see if it is curved or flat. The generalization from a two-dimensional curved surface to the four-dimensional curved surface of spacetime is a challenge to the imagination but not necessarily to the intellect. We may not be able to draw or picture the four-dimensional space, but we can easily understand what it means to be non-Euclidean in four dimensions. A geodesic in two dimensions is the straightest possible path between two points. A geodesic in three dimensions is the straightest possible path between two points. A geodesic in four dimensions is the straightest possible path between two points. Light, being the fastest possible signal between two events, must follow a geodesic through space and time. Since we are talking about the *speed* of light, and not just the *distance* it travels, we must consider not just the spatial dimensions but also the dimension of time. As a somewhat technical note, realize that rays of light are not the only geodesics in spacetime. They are the geodesics that connect events that have a lightlike separation. Events that have a timelike separation are connected by geodesics that correspond to things traveling at less than the speed of light.

Light follows geodesics in spacetime; that is what is important here. This should be no surprise. Surveyors often use laser beams of light to mark the straight line between two points or to triangulate on the position of a third point. Common sense and the special theory of relativity agree that light travels in a straight line, or in the more generalized language we are now using, on a geodesic. But we have already seen that a strong gravitational force causes light rays to bend. This is both a conceptual conclusion, as the effect is a consequence of the principle of equivalence, and an experimental conclusion, as it was observed during the solar eclipse. A massive object like the sun bends the nearby geodesics.

As an attempt to visualize this situation, Figure 4.7 compares the behavior of two geodesics on the curved surface of the earth to the behavior of two rays of light passing near the sun. The picture of the light bent

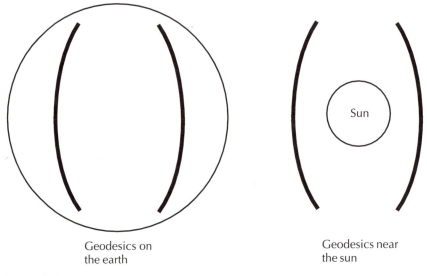

Geodesics on
the earth

Geodesics near
the sun

Figure 4.7

by the sun is a very stylized, exaggerated two-dimensional rendition of events that take place in the four dimensions of spacetime. Nonetheless, the similarity between the two cases of curved geodesics is noteworthy. Geodesics bend in order to follow the curvature of the space in which they are embedded. Thus, the bending of light rays is itself evidence of the curvature of space and time.

The principle of equivalence has thus brought us to an account of gravity in terms of the curvature of spacetime. Massive objects like the sun or planets alter the geometry of the spacetime around them. They cause the geodesics to bend in toward the center of the massive object. Thus, things moving through spacetime, things like light or any kind of object, will curve along with the spacetime and move toward the massive source. The attraction of gravity, things like the ball falling to the floor or the moon staying in orbit, happen because the spacetime is curved in toward the source. Massive objects cause a curvature of the spacetime, and massive objects respond to a curvature of the spacetime. Gravity, the force of attraction between massive objects, is entirely mediated by the geometric structure of spacetime.

MACH'S PRINCIPLE

What can be said about the substantival or relational nature of space and time, in light of this discussion of gravity and curvature? In the general theory of relativity, spacetime has dynamic properties. It is curved by the presence of massive objects. The more massive the object, the sharper the curvature. But this in itself does not mean that the spacetime of the general theory of relativity is substantival. Relations can have dynamic properties, and a curved spacetime can be a relational spacetime.

An analogy can help clarify the point. Consider the relation of marriage. Marriage is not a thing in itself, a substantival container waiting to be filled by pairs of people. It is just a relation between people. Without the people there is no marriage. This is a relational account of marriage. But even as nothing more than a relation, a marriage can take on dynamic properties. A marriage can be loving and warm, or it can be cold and hostile. It can be beneficial or abusive. It can be short-lived or long-lasting. And so on. These properties are brought about by the participants themselves, and these properties then affect the participants. The people create hostility and they respond to hostility. The situation is much like massive objects bringing about the geometric properties of spacetime, geometric properties like curvature that also affect the massive objects. The objects create curvature and they respond to curvature. The point of the analogy is that ascribing properties to something like marriage or spacetime does not presuppose that it is a substantival thing with an independent existence. For the spacetime of the general theory of relativity then, the question is still open.

The key to answering the substantival/relational question is whether the spacetime has any properties independent of the objects in it. The defining tenet of a relational model of spacetime is **Mach's Principle**:

> All properties of spacetime are caused by massive physical objects, and all spatial and temporal properties of things are in relation to other things and events.

At first glance, the spacetime of the general theory of relativity seems to fit Mach's principle nicely. Physical objects cause the curvature of spacetime and physical objects respond in observable ways to curvature. The spacetime and its curvature seem to be just a way to talk about the relation between objects that we measure as gravitational attraction and acceleration.

But a closer look at the nature of spacetime in the general theory of relativity shows that it does in fact have well-defined, determinate properties of its own that are not caused by, nor defined in relation to, any physical object. If the sun were not there, then the rays of light from distant stars would go straight. Geodesics would be parallel. Spacetime, in other words, would be flat. The presence of a massive object like the sun does not define the properties of spacetime, it modifies them. Using the theory in this way indicates that spacetime is flat except where massive objects alter this antecedent flatness. Like bowling balls on a trampoline, there is an intrinsic flatness that gets modified. Far from any bowling balls, the surface is flat. If there were no bowling balls at all, the surface would be flat. Similarly, in the language of general relativity, spacetime is "flat at infinity." This sort of boundary condition is necessary for deriving determinate solutions to the equations relating the physical objects that are sources of gravity and the curvature they produce. The assumption of flat spacetime infinitely far from the source, or some other specification of the properties of spacetime far away, has to be added to solve the equations. Mach's principle would require that infinitely far from any physical object spacetime would have no properties at all. Its curvature would be indeterminate, not flat. Furthermore, the general theory of relativity allows us to talk about properties of spacetime in a universe with no objects at all. Mach's principle would again require that such a thoroughly empty spacetime be entirely without features, but in the general theory of relativity the empty spacetime has geometric properties.

Even with this decidedly non-Machian aspect of the general theory of relativity, it would be misleading to conclude that the issue of substantival/relational spacetime is settled. Other implications of the theory pose challenges to substantivalism. Recently renewed interest in an argument first proposed by Einstein, his so-called hole argument, suggests that understanding the nature of spacetime as it functions in the general theory of relativity is a work in progress. In fact, our concepts of substantivalism and relationalism might be works in progress in the sense that the challenges of the hole argument may force revisions in our understanding of these ideas. It should not be a surprise, or a disappointment, to find unresolved issues in the newest theories of science.

Before outlining the hole argument it is important to note that the general theory of relativity is a very non-experimental branch of physics. Unlike the special theory, where the theorized effects such as time dilation and length contraction can be easily created in the laboratory and measured in a large variety of phenomena, there are only a few measurable effects of the general theory of relativity. There is the gravitational redshift, the bending of star light around the sun, and a small influence on the orbit of the planet Mercury. In general, the strength of gravity required for measurable relativistic effects is enormous, too large to produce in a laboratory. Most of the work with the general theory of relativity is therefore theoretical in the sense of finding consistent models of spacetime that are allowed by the theory. Under specific, hypothetical situations of mass distributions, what is the resulting spacetime structure? And what will happen over time? These sorts of questions and the resulting calculations can lead to interesting and perhaps observable results. It is this approach, for example, that generates the idea of a black hole.

The theoretical model that is troublesome for a substantival account of spacetime is in Einstein's hole argument. Imagine a section of spacetime, large or small, that is entirely empty. This is the hole, a region with no matter at all. The rest of spacetime, the surrounding, does contain matter and so its geometrical features are determined by the distribution of mass. But what is the geometry of spacetime in the hole? The formalism of the general theory of relativity proves that the masses and spacetime structure outside the hole do not determine a unique geometric structure inside the hole. That is, there is any number of different geometric structures in the hole that are consistent with the geometry on the outside. We should note that the spacetime of the general theory of relativity is not the only one that allows for such indeterminate holes. A broad class of theories leads to the dilemma.

These results are usually presented as an incompatibility between the substantival account of spacetime and the metaphysical idea of determinism. It amounts to a serious challenge for substantivalism. According to the substantivalist, there is still something in the hole, namely the spacetime. It is real, but it has no determinate properties, at least no determinate geometric properties.

The hole is not such a problem for a relational account of spacetime. In this model, the hole is genuinely empty, and the different possibilities of geometric descriptions allowed by the general theory of relativity and the surrounding spacetime structure are just alternative descriptions of the same physical situation. There is no indeterminate reality in the hole, no metaphysical indeterminateness, because there is no real thing in the hole.

The logic of the hole argument's challenge to substantivalism is very much like Leibniz's argument at the beginning of Chapter 3. The substantival model of space, he argued, allows any number of different physical situations that are all consistent with what we observe of the distribution of objects in the universe. A universe in which all objects were shifted to the left by 1 foot would be indistinguishable from the unshifted universe. So, whether substantival space has everything like this (not shifted) or like that (shifted) is not determined by what we can observe. Nothing determines how things must be located in space, thus an essential property of substantival space is indeterminate. Similarly for the hole in the general theory of relativity, nothing determines the essential geometric properties of the spacetime.

The hole argument, and other reflections on the character of spacetime as described by the general theory of relativity, does not lead to a definitive rejection or acceptance of either the substantival or relational model. They force, first of all, a refinement of the metaphysical concepts, including the concept of determinism. This is an ongoing project these days, and it is worth noting that the intent of the project is to determine, in the context of the general theory of relativity, the nature of spacetime. The intent, in other words, is to figure out the reality, the real nature of spacetime according to the relevant theory of physics. There is no suggestion of settling for the appearance of spacetime.

SUMMARY OF THE GENERAL
THEORY OF RELATIVITY

The principle of general covariance demands a new account of the nature of gravity, just as the principle of relativity forces a new account of the speed of light. Both required changes are unlikely and surprising, and both present profound challenges to common sense. Why would the principle of relativity affect our understanding of *light*, of all things? Because light, or better, electro-magnetic radiation in general, is unique in that its speed of propagation is an explicit part of the laws of electricity and magnetism. But then why *gravity*? Because it is a universal force that affects all matter in the same way, regardless of its composition. This universal nature of gravity means that its effects are equivalent to the effects of acceleration. Because of this equivalence, the objects that are the source of gravity are what determine the properties of motion of things in the universe.

The hope of a theory of relativity was to avoid any reference to spacetime itself as an irreducible determining factor in the laws of physics. The principle of general covariance would have us do entirely without a special, singled-out reference frame, such as the reference frame of spacetime itself. Covariance wants all reference frames to be equal, so that the laws of physics cannot recognize one from the other. It requires that the laws not recognize a thing that is spacetime itself that entails a reference frame.

In the general theory of relativity, *inertial* reference frames are determined with respect to the geodesics of spacetime, in the following way. Something that is moving on a geodesic has no acceleration. It is not turning in spacetime, nor speeding up, nor slowing down. It is maintaining a steady course through space and time. This is what we mean by the generalized notion of a straight line, a geodesic. With no acceleration, following a geodesic is thus exactly what we expect of an inertial reference frame. In the flat spacetime of a Minkowski diagram, the spacetime appropriate to the special theory of relativity, the geodesics are in fact straight as drawn on the Minkowski diagram. Thus any trajectory that follows a straight line, any straight line, is suitable for determining an inertial reference frame. The straight lines of the diagram, the class of all the straight lines, determine which frames are inertial and which are not. Generalizing to any spacetime, flat or curved, the geodesics determine which frames are inertial. Following a geodesic means moving with no acceleration, and that is the criterion of an inertial frame.

The geodesics of a spacetime are determined by the massive, physical objects in it. This is the part of the theory that is relativistic, that gets us close to avoiding all reference to intrinsic properties of spacetime itself. Recall that the presence of the sun altered the shape of the geodesics nearby from being straight to being slightly curved. Thus the determination of inertial reference frames is ultimately by reference to objects. Space-time does not have the inertial properties itself (except for the antecedent flatness discussed in the previous section).

If the reference to spacetime is just a convenient way to talk about the effects of one object on the motion of others, then we ought to be able to eliminate reference to spacetime altogether. Toward this end, we can say that a reference frame that is in gravitational free-fall is an inertial reference frame. Think about the light that bends as it passes the sun. The light itself is in free-fall. There is no force acting on it other than gravity. If the effects of gravity are built into the shape of spacetime, then there are no forces acting on the light and it will follow a no-acceleration trajectory, a geodesic. In a reference frame that is free-falling toward the sun, in a box that is just let go to fall into the sun, the ray of light passing by from a distant star would look straight. That is, by falling toward the sun with it, the bending is canceled out. The path of the light, the geodesic, is straight and you are back in an inertial reference frame of a flat space-time where you can do physics on a Minkowski diagram and use the special theory of relativity.

Gravity is truly ubiquitous throughout the universe, and it complicates our lives by curving the geodesics of the surrounding spacetime. But you can always cancel out the effects of gravity by falling freely into the source. Give yourself up to gravity and you will not be troubled by its force or its alteration of the spacetime. Your physics will be inertial and your spacetime flat. But for those situations where we cannot fall freely, for example when the surface of the earth is pushing up on us to prevent it, we will still need to talk about non-inertial reference frames and curved geodesics.

Chapter 5
RELATIVITY AND REALISM

Bohr's claim about the limitations of physics, and the inability to know how nature really is, was motivated by quantum mechanics. There is something about the quantum mechanical description of nature that suggests and, for Bohr and those who share his interpretation, justifies the anti-realism. We will get to quantum mechanics and see what it is about that theory that provokes the idea that all we can know are appearances. But first we should see if the theory of relativity, the special theory of relativity, and the general theory of relativity together is in any way as provocative. The question for this chapter is whether Bohr's claim is in any way suggested or supported by the details of relativity. Perhaps the most evenhanded, and the most informative way to put it is this: In what sense is relativity about how nature *is*, and in what sense is it only about how nature *appears*? To what extent is relativity about a reality that is independent of us and our way of looking at things, and to what extent is it about us and our interactions with nature?

TWO SEPARATE QUESTIONS

It is important to recognize and separate two different questions involved here: What is relativity about? and How do we know the theory of relativity is true? We will address both issues, but it is the first, the one designed to reveal what physics itself says, that is our main concern.

The second question is a classic question of epistemology. It returns us to the philosophical arguments about scientific realism. What can be proven, and how? Does the evidence, the appearances, justify the conclusion? The first question also has an epistemological flavor to it. As we will pursue it, the question is what the theory itself indicates about our ability to know about nature. This will involve both what the theory claims explicitly and what it implies about the information that is available to an observer. The proof of the theory aside, we can ask about what the theory says and what it intends to describe.

The distinction between these two separate questions is similar to the distinction between the physical and conceptual influence people have on the content of their knowledge. The physical influence, the fact that we can observe and know about the world only by physically intervening with the objects and events of interest, is accountable to a physical science such as physics. Physics is, in part, about the details of this kind of intervention. We turn to physics to understand the physical interaction and see if, in a particular case, it can be avoided or transcended. Look to the theories and experiments of physics to see if there is always a human component, an effect of physical interaction, in the results. Thus, the question of what physics is about and whether humans are always physically in the picture is a question about the physical influence on knowledge. The question about proof, about the justification for the theory, is more about our conceptual influence. It is asking about preconceptions that influence the evidence and the logic of confirmation. Perhaps old beliefs or the theory itself enter into the proof, putting it on implausible or untested foundations. This is, of course, an issue of crucial importance. It would be irresponsible and incomplete to ignore this second question. So we will not. We will deal with the two questions separately, in turn.

THE WAY NATURE IS

For the moment, put aside the question of how we know the theory of relativity is true. Concentrate on the more internal issue. Does relativity describe how nature is, or only how nature appears to us? It describes how nature is. The intent of the theory, or, to use Bohr's word, the task of the theory, is to tell us the way nature is.

To see that this is the case, and that the theory of relativity gives no support for Bohr's claim, start with the special theory of relativity. A very superficial reading of the special theory of relativity might make it seem like just the thing to make Bohr's point. Being superficial to the extreme of only attending to the name of the theory, someone might summarize by saying that Einstein's theory says that everything is relative. Expanding on this misconception, there are no absolutes, no one way that things are, no single truth of any matter. The way nature is is just relative to each observer. My take on nature will be different from yours, and there is no saying that one is right and the other wrong. All we can talk about is a person's relation to nature. It is all relative.

I may have built a straw man here. I hope I have. But it will serve a useful expository purpose. The claim that Einstein's theory says that everything is relative is a grossly misguided paraphrase of the theory. As we have seen, the special theory of relativity is a theory built on absolutes. The conceptual foundations of the theory are the absolute nature of the laws of physics and the absolute speed of light. The kernel of the theory, in other words, is about what is *not* relative. Furthermore, many of the principles that follow from these absolute foundations are about absolute properties. We did not make a big deal about the properties that are not relative in the special theory of relativity, only because they do not play a distinctive role in the theory. That is, they do not show off what distinguishes relativity from Newtonian physics. But the electric charge of an object, for example, is not a relative property. It is invariant, the same in all inertial reference frames. A property called the spacetime interval is also invariant. In the Newtonian description of mechanics, the spatial interval, the length between points in space, is invariant. In the relativistic account, the space*time* interval, the four-dimensional distance through space and time between events, is invariant. Every inertial reference frame measures the same value.

Clearly, not everything is relative.

There is an important lesson to be learned from the straw man, a lesson to be remembered when we get to quantum mechanics. A demonstration that a few properties are relative does not show that *all* properties are relative. The special theory of relativity has to analyze properties on an individual, case-by-case basis, and some come out to be relative but others come out to be absolute. There is no denying that the ones that are relative, properties like time duration, simultaneity, and length, are genuinely amazing. But we cannot let our amazement at these overwhelm our good judgment and prevent a logical assessment of the theory. Proving that some things are relative is not proving that all are.

Furthermore, what about the features that are relative? Are they relative to us, to human observers? In other words, does the relativity of length or time duration make our knowledge of such things subjective in the way implied by Bohr? No. According to the special theory of relativity, the phenomenon of length contraction, time dilation, and the like, happen even when no person is present and when no one is looking. These are not the effects of interaction by an observer. The properties that are relative are relative to a reference frame, not to a person or to an act or manner of observation. They are relative to a state of motion.

An effect such as length contraction is not caused by the act of measuring length. It is not like the effect of a thermometer heating up the sample. Nor is it like a foreshortening effect in which the distant mountain looks small enough to crush between your fingers. Neither of these effects happen unless someone is looking or at least performing the measurement. The foreshortening and the raised temperature are properties of the image, not of the object by itself.

Length contraction, and all other special relativistic effects, happen when no one looks or measures. The lifetime of muons, for example, is extended when they are moving at the fantastic speeds as cosmic rays. They travel a certain, determinate distance through our atmosphere and, in light of the special theory of relativity, have always done so, even before we knew about them or could detect them. The description of how long they exist, how fast they are going, and how far they travel does not include us or our way of looking at all. It is a description of what is going on in nature.

The relativity of simultaneity is another clear case in which the relativity is not relative to an observer or an act of measurement. It is not a subjective relativity to a person. It is not that two events *appear* different to different people, as the lightening and thunder appear at different times. Rather, *being* simultaneous is relative.

Another good example of the objective nature of special relativistic effects is the use of the theory to build machines such as particle accelerators. The Stanford linear accelerator, for example, moves electrons along a 2-mile path and accelerates them to almost the speed of light. At such speed, relativistic effects are large and must be reckoned with. One property we have not discussed is mass. It is a relative property. Opposite of the behavior of length, which decreases when an object moves, mass increases. Mass increases not just when an object is observed to move, but when it moves, observed or not. So the many interactions between an electron and the alternating electric fields along the 2-mile course of the accelerator must be timed to interact with an even more massive electron. Nobody observes these interactions. They are, according to the special theory of relativity, happening between mindless, objective, independent entities such as electrons and electro-magnetic fields. These interactions themselves are relativistic. That is the way things are in nature.

One of the key accomplishments of the special theory of relativity is to clarify the nature of relative properties like length and mass. These properties are determinate only with respect to a state of motion. They are two-place properties, in the sense that they require two pieces of information to be fully specified. You cannot say simply that something is 10 meters long. The 10 meters is just one piece of information. You have to say it is 10 meters long with respect to the platform reference frame. Specifying the reference frame is the second piece of information. Things *have* length only with respect to a specific reference frame. This complication, this relativity, is not in how you look at things; it is in how things are.

We deal with two-place properties all the time in simpler and more common situations than the special theory of relativity. The property of being taller than is clearly two-place. So is the property of being inside of. You cannot say simply that Mont Blanc is taller. This lacks the second piece of information, taller than what? It is taller than the Matterhorn. Taller than is a relative property, but it is entirely objective, because the second component, what it is relative to, is a determinate thing in the world, not just some personal matter of taste or perspective. It is relative to another thing, in this case a mountain. The point is that something is or is not taller than only in relation to some object of reference, but there is a fact of the matter and we can know it. The special theory of relativity indicates that, contrary to initial intuition, length and mass and time duration are also relative. Things have length and mass and so on, only in relation to some frame of reference. But here again there is a fact of the matter as to the length or mass of something, given the frame of reference.

This is what, according to the special theory of relativity, we can say about length and the other relative properties. Getting back to Bohr's words, physics tells us what we can say about nature. But in this case, physics is stricter than that. The special theory of relativity tells us what we *cannot* say about nature. We cannot say that length or simultaneity is an absolute property. We cannot say that the speed of light is relative. We cannot say these things because they would lead to an inconsistency in our account of nature, and an inconsistent model must be inaccurate in some way. Some part of the description must be false, that is, not true of nature. We cannot say that the speed of light is relative simply because it is not relative. We have no choice. In the case of the special theory of relativity then, it is more than just what we *can* say of nature; it is what we *must* say.

This interpretation of the special theory of relativity focuses on the explicitly relative or absolute properties of things. These typically involve separated points in spacetime. The length, for example, refers to two separate end points in space, and of course, time duration involves separate points in time. It is also worth asking about properties at a single point in spacetime, that is, at a single event. Mitch's paradox makes a good example to illustrate the objective nature of events in the special theory of relativity.

We can look at Mitch's paradox, not to revive any of the challenges to our understanding of the special theory of relativity, but simply to point out some of the features of the setup itself. At the event of their meeting, Richard is younger than Bubba. It is not that he seems or appears younger to one of them or the other. He *is* younger, period. We do not have to ask, how does Richard appear to Bubba, or how does Bubba appear to Richard. It is the same result for either observer, or any third observer. By any test, a group photograph, counting wrinkles or gray hair or dead brain cells, Richard is younger than Bubba.

We could do Mitch's paradox entirely without people, in case Richard and Bubba are giving you a feeling of subjectivity. We could do it with piles of decaying, radioactive atoms, two of them sitting on opposite ends of a lab table, the third moving by the table. The two on the table are the same age in the sense that, in the lab reference frame, they start with the same number of radioactive atoms. When the third flies by the first, they have the same number of radioactive atoms; they are the same age. When the third flies by the second, the third has more of the radioactive atoms (and fewer of the products of the radioactive decay) than the second. The third has aged less. This is a fact of the event that is not altered or influenced by observation or by people in any way. This is not a description of how we interact with the world, it is how the world behaves. Furthermore, the special theory of relativity explains why Richard is younger (or the third pile has more of the original radioactive atoms left) in entirely objective terms. It has to do with the simultaneity of the events of the birth of Bob and Bubba. They were born at the same time in the lab reference frame, but not in the reference frame that is moving with respect to the lab.

Because of its explicit recognition of the link between reference frames and the values of measurements, the special theory of relativity is particularly well suited to facilitating objective knowledge. Objective knowledge is achieved by recognizing the effects of one's own perspective. Since all measurements must be done in one reference frame or another, the relativistic effects of different reference frames are also effects on our observations and measurements. Knowing the facts about these effects will help in reconstructing the link between appearance and reality. Every informational signal comes to us at a finite speed, less than or equal to the speed of light. Simultaneity is relative to reference frames. So are length, time duration, and other properties. Now that we know this, we can factor this information into our account of the natural world. Scientific observation is observation with all things considered. The special theory of relativity is a large part of what needs to be considered.

What about the general theory of relativity? If we change the discussion from properties of things like length or simultaneity to curvature and gravitational red shift, are we still describing the world or is it our way of interacting with the world? This question returns us to an unresolved issue in Chapter 4. As a preface to the description of curvature, we asked whether the curvature was a property of the natural thing being measured, or more properly of the process of measurement.

An effective way to answer this is to review the reasoning that leads from the special theory of relativity to the general theory of relativity. A summary account of the special theory of relativity would say that motion affects some of the properties of things. The step to the general theory of relativity expands on this only by saying that it is not only unaccelerated, inertial motion that affects properties; accelerated motion does as well. It is the same kind of effect in that it is not caused by the process of measurement. It is a feature of the lengths and time durations themselves, just as special relativistic effects are. The general theory of relativity then links these sorts of effects to gravity and the material objects that are the source of gravity by the principle of equivalence. This step reveals an asymmetry in the general relativistic effects that is not present in the special theory of relativity. There is a definitive nearer-to and further-from a source of gravity, and gravitational effects will reflect this orientation. Clocks nearer to the source run slower than clocks further from the source.

The consequences of the equivalence principle are descriptions of natural occurrences, not of interactions with people. Light bends around the sun whether we are measuring it or not. And, perhaps the most remarkable prediction of the general theory of relativity, black holes exist whether we observe them or not. The theory claims that they are out there, undetected. The features of a black hole, for example that its intense gravitational force red-sifts light to a frequency of zero and hence right out of existence, are features of the things themselves, not of how we look at them.

The general theory of relativity, like the special theory of relativity, is about nature. Its task, its intent, is to describe how the world is, not merely how it appears to us. The theory of relativity as a whole neither suggests nor justifies Bohr's assessment of physics. It certainly does not imply or assume that there is no objective, independent physical reality out there for physics to describe. Nor does the theory of relativity show that the picture provided by physics must always include a component of human interaction. By explaining what parts of the appearances are the effects of our influence and perspective, the theory of relativity allows us to erase these from the final picture. Thus, the theory certainly does not indicate that there is no getting beyond appearance to an independent reality.

To summarize, the theory of relativity itself does not entail a limitation of knowledge to mere appearance. There is nothing in the theory that indicates we cannot know how nature really is.

This summary ignores the central epistemological question of how we know the theory is true, and we can now ask whether a broader look at the theory, considering not just its content but its proof, suggests Bohr's point about the limitations of physics. This aspect of the issue will be particularly important for the general theory of relativity since its description of nature goes a conceptual step beyond the special theory of relativity. The special theory of relativity describes properties of things that we can observe and measure, things like the length of a railroad car or how much time it takes for a bug to eat a pencil eraser. But the general theory of relativity describes properties of something we cannot observe, the curvature of spacetime. This extra step of abstraction raises the epistemological question, How do we know the theory is true?

HOW DO WE KNOW THE THEORY IS TRUE?

The question now is about the plausibility of the link between the evidence for a theory and its descriptive model of reality. We can start, as usual, with the special theory of relativity, this time because it is the quicker and easier case.

The special theory of relativity is, in the context of science in general, a very well-confirmed theory. There is a huge variety of experiments done that give results in favor of the special theory of relativity. The cosmic-ray muons are only one example. The theory not only makes predictions (such as the one about muons) that turn out to be true, it is also used with virtually all other aspects of physics to design and interpret experiments. This shows that the claims of the special theory of relativity are not narrowly self-serving in the sense of prescribing and interpreting their own observations. The relativistic effects are essential in designing much of the machinery of physics, like the Stanford linear accelerator. If the special theory of relativity were not factored into the design of these machines, they simply would not work. The special theory of relativity is also essentially involved in other theoretical models of physics and astronomy. Our understanding of the fine structure and the behavior of atoms includes the special relativistic effects. The atomic physics then serves astrophysics in helping to understand the nature of processes observed in distant stars and galaxies. The special theory of relativity, in other words, plays a ubiquitous role in a diverse and coherent description of nature.

All this is meant to show that, if any scientific theory is justified by the evidence, the special theory of relativity is. This, of course, does nothing to meet the general challenge from the anti-realist philosophers who claim that *no* theory is justified by the evidence. Insofar as our project is to reveal what the physics itself indicates about our ability to know nature, the more general philosophical debate is not really our business. Nonetheless, I cannot help but make a few quick comments on the justification of scientific theories. The anti-realist indicates that in general there is no plausible inference from appearance to a reality beyond, no reasoning from the evidence that things are *as if* some theory is true to the conclusion that the theory *is* true. There is no certainty, no foolproof justification, and so, in the end, belief that a theory is true can be no more than faith. But it is important for us to acknowledge that there is some room to maneuver between irrefutable, certain proof, on the one hand, and blind faith, on the other. Surely there can be degrees of justification in the sense that I have more reason to believe that the sun will rise tomorrow in the east than I do to believe that at the core of the sun, hydrogen is fusing together to form helium. Neither claim is certain, and neither is an act of faith. The interesting difference between them is the difference in degree of proof, of justification. We would toss out the whole analysis, and thereby hide from our responsibility to judge plausible from implausible claims, if we said simply that what is not certain is faith. What we need to know about the special theory of relativity and other scientific claims is how much justification they have.

There is a contradiction that threatens the skeptical position, not an inevitable contradiction, but one to be advised of. If there is some epistemic impediment such that we cannot know that nature really is as the special theory of relativity describes it, then the same impediment precludes our knowing that nature is *not* as described by the special theory of relativity. A definitive claim that the special theory of relativity is not about how nature is, is a claim that violates the very skeptical limitations it is meant to support. The tenable skeptical position must be one of agnosticism. If there is really a general limit to our knowledge, and the way nature is (or is not) is beyond it, then we must plead ignorance in response to the question of whether the special theory of relativity really describes the way things are. We cannot know, one way or the other.

But here we ought to recall the analysis of the special theory of relativity, showing that the theory dictates not just what we can say about nature, but what we must say. Why is it that we must say that the speed of light is absolute? Probably because it *is* absolute. This is the link between appearance and reality, a degree of justification.

The general theory of relativity though, is on thinner ice. It is not among those you would call a very well confirmed theory. There are not lots of experiments whose results shed light and plausibility on the theory. There are no experiments that tell against the theory, but there are relatively few that are for it either.

The fact that spacetime appears to be curved does not necessarily mean that spacetime is curved. We need to distinguish between the claim that measurements of things like the trajectory of a ray of light are non-Euclidean, and the claim that spacetime itself is non-Euclidean. The general theory of relativity takes the extra metaphysical step, from describing the path of a beam of light to describing the nature of spacetime. In deciding whether this step is warranted, we need to realize that there are alternative conclusions that can be reached from the same basic evidence. In this case, physics gives us a choice on what we can say about nature, and the details of the physics lead to a demonstration of underdetermination of theory by evidence.

Our access to information about curvature and the geometry of spacetime is by measuring spatial distances and durations of time. Measure the diameter and the circumference of a circle, for example, and if the ratio of the measurements is not π, then conclude that the space is not flat, not Euclidean. Measure the three angles of a triangle and if their sum is not 180°, draw the same conclusion. But the measurements could be interpreted differently. Perhaps it is the measuring tools that are changing rather than the thing being measured that is not flat. In other words, the space may be perfectly flat, perfectly Euclidean, but, because of some invisible force, the measuring device expands or shrinks depending on where it is. This would account for the results that the circumference is not measured to be π times the measurement of the diameter.

On this interpretation, the explanation of the non-Euclidean measurement does not invoke the counter-intuitive and otherwise unobservable entity of non-Euclidean spacetime. It does, however, rely on a new and otherwise unobservable force, introduced entirely ad hoc to accommodate the curious measurements. It would have to be a ubiquitous and egalitarian force that affects all objects in the same way and to the same degree, just as gravity does. This universality is essential because all techniques of measurement must give the same results. We do not suffer the ambiguity of having different values for length-by-a-wooden-ruler, length-by-a-steel-ruler, length-by-lightbeam, and so on. The force that distorts measuring devices must affect all sorts of things in an identical way. This new interpretation, in other words, introduces in an ad hoc way, a universal and otherwise undetectable force.

There is nothing necessarily wrong with an ad hoc idea. Some great ideas in science have shown up ad hoc and gone on to play credible, useful roles in successful theories. Galileo's concept of inertia is a good example. The idea that objects continue to move on their own inertia, even when no external force is helping them along, was introduced in part to explain away evidence that was otherwise contradictory to the Copernican and Galilean theory that the earth rotates on its axis. An ad hoc idea should not be immediately embraced, but neither should it be automatically dismissed. A decision has to be made as to which is more plausible, the universal force that alters the size of measuring devices in a flat space, or a curved space in which measuring devices maintain their size.

A simple example of measurements confined to a two-dimensional surface will help clarify the similarities and differences in the two models of spacetime. The important point is that they agree on all the evidence. They disagree on what the evidence means. Put another way, they agree on everything that is or can be observed. They disagree on claims about what cannot be observed.

It is important in the example to specify not just what to measure but how to measure it. This will ensure agreement on the basic data of observation. Start by drawing a big circle. A good way to do this is to use a string that is fixed at the center of the circle and that swings around to mark off the set of points that are all the same distance (one string-length) from the center. This set of points is a circle. Now measure the circumference, c, and the diameter, d, by laying a unit measuring stick from end to end along each distance. Suppose we find that the measurements of c and d are related as $c < \pi d$. Because of this inequality, the measurements are non-Euclidean. All of this, the setup and the results of the measurements, is agreeable to both models of spacetime. They disagree on what to make of it. Under the influence of the general theory of relativity we would conclude that, since non-Euclidean measurements happen on curved, non-Euclidean surfaces, the experiment must have been done on a curved surface. But the alternative interpretation points out that non-Euclidean measurements can happen on a Euclidean surface if measuring tools shrink or expand in an undetectable way. Suppose, the alternative suggests, the circle was drawn on a huge hot plate that was cold at the center and got hotter closer to the edge. The measuring stick would expand when it was out measuring the circumference, thereby requiring fewer layings end-to-end to get all the way around the circle. This would account for the low measured value of c and the fact that $c < \pi d$. The measurement shows that the circle is on a flat surface but some factor is at work that expands measuring sticks as they get further from the center.

Of course, the idea of a giant hot plate is just a fanciful heuristic tool. We cannot really say that heat is the cause of expansion, both because we could feel the heat if it were there (that is, heat has other effects, in addition to expansion), and because heat expands different materials by different amounts. A steel measuring stick would get longer relative to a glass measuring stick and, by comparing the two, we could detect the expansion. The argument has to be made with some other kind of force, some new force that is thoroughly universal, otherwise the claim that measuring sticks expand can be directly checked.

So is the surface flat or curved? The observations, the measurements, cannot answer that question. The evidence is neutral with respect to the geometry of the surface itself. The evidence does not determine whether it is flat or curved. If we assume from the start that there are no universal forces that affect measuring sticks, then the evidence shows that the surface is curved. But if we assume that the surface is Euclidean, then this same evidence shows that measuring sticks are changing length under the influence of some unseen universal force. The description of the geometry of the surface depends on the choice of assumption we make at the beginning. The geometry is, in this sense, a matter of convention.

The same conventionality is true of geometry in three-dimensional space or four-dimensional spacetime. Without an initial assumption about the behavior of measuring tools there is no way to infer from the evidence of measurements to a conclusion about the geometry of spacetime. Whether we call it curved or flat ultimately depends on what convention we adopt to account for measuring tools.

There is a similar kind of conventionality in the special theory of relativity. I have put off mentioning it until now in order to deal with the whole issue of conventionality at one time and to assess its implications for our ability to know what nature is like. In the general theory of relativity, the geometry of spacetime depends on our choice of a measurement convention. Nature does not tell us which choice to make, so nature doesn't give us enough information to specify the geometry. In the special theory of relativity, within a single reference frame, the simultaneity of two distant events depends on a convention, a choice, about the one-way speed of light. The result is known as the conventionality of simultaneity.

This is not the *relativity* of simultaneity we are talking about. The relativity of simultaneity describes the differences of simultaneity according to different reference frames. The conventionality is an issue within a single reference frame. Restricting our attention to just one reference frame we can ask whether two distant events are simultaneous. Strictly on the train car, for example, are two flashes of light, one at each end of the car, simultaneous? Answering the question will always depend on noting the arrival of the two flashes of light at one point, the mid-point, for example, and then calculating how long it took the light to get from each event to that point. The two signals of light from the two events travel in two different directions. Since the calculation requires knowing the speed of light, it requires knowing the speed in each of the two directions. In our examples we have always assumed that the speed is the same in the two directions, but this turns out to be exactly that, an assumption. The special theory of relativity *requires* certain properties of light. The speed is absolute, that is, the same in all reference frames. The roundtrip speed, from here to there and back to here, is always c. The speed is independent of the speed of the source. No causal signal can go faster than the speed of light. But there is no conceptual requirement that the speed of light in a particular reference frame is the same in all directions. All possible measurements of the speed of light are either measurements of the roundtrip speed or measurements of one-way speed using clocks at each end that have been synchronized in terms of simultaneous events that are determined by knowing the one-way speed of light. There is no measurement that directly reveals the one-way speed of light.

The details of the special theory of relativity, principles like the finite speed of light and the nature of measurement, reveal that the one-way speed of light cannot be determined. In all of our examples in Chapter 3 we assumed the speed of light was them same in all directions. But this was a choice, a matter of convention like the convention on the nature of measuring sticks that they do not shrink or expand as they move around the universe. No experiment could disprove an assumption that the light goes faster one way than another, as long as it is consistent with the roundtrip speed of c. The results of the special theory of relativity, measurable phenomena such as time dilation and length contraction, are substantially independent of this element of choice. That is, different choices of the one-way speed of light all result in time dilation and length contraction. The properties of time duration and length are relative properties, even though simultaneity within a single reference frame is a matter of convention.

This shows that an element of conventionality in the theory does not make its description of nature entirely a matter of choice. Conventionality is not a sign of subjectivity. There are still facts of nature that emerge. Is the simultaneity of two distant events relative to the reference frame? Yes, period. There is no conventionality, no choice, no subjectivity in this. It is not a matter of appearance; it's the way things are. Are the two events simultaneous within a single reference frame? Well, this depends on the one-way speed of light convention. But once a particular convention is adopted, there is a right answer. It is not that the special theory of relativity indicates that there is some aspect of nature we cannot know. It is rather that this property is actually indeterminate in nature. It is not really there, but if we insist on talking about it we will have to choose one value or another. And whatever choice we make, we end up with much the same description of what is determinate in nature.

The story in the special theory of relativity is that one aspect of the theory indicates that another aspect is indeterminate and a matter of convention. The same is true in the general theory of relativity. To see this requires clarifying the difference between geometry and topology. Think of the properties associated with a surface like the hot plate or a higher dimensional space. It can be flat or curved. It can be of infinite extent or bounded. It has a certain number of dimensions, two for the hot plate, four for spacetime. It can be smooth and continuous or discrete in a way that there are gaps between the points of space. And so on. Some of these properties require distance measurements for their determination, some do not. The number of dimensions, for example, is independent of the distances between points. So is the property of continuity or discreteness, or the property of being bounded or not. These are topological properties. Contrast these with geometrical properties that do depend on determination of distances.

Curvature, we have seen, is closely related to distances like circumferences and diameters. Curvature is a geometric property of a space. So is the property of being finite or infinite. An amusing and trustworthy way to visualize the difference between properties that are geometrical and those that are topological is to image the space as a stretchable sheet, like a trampoline. Any property of the sheet that changes when the sheet itself is stretched is a geometrical property. Properties that can only be changed by actually cutting or gluing together pieces of the sheet, are topological properties. Putting a bowling ball on the trampoline stretches part of it. This curvature is a change in the geometry, but the topology of the sheet is unchanged. If you looped the sheet around to make a cylinder, joining two opposite edges together, you change the topology from being bounded on those two edges to being unbounded. A bug could walk forever in the looped direction and never reach a boundary of the sheet.

Some topological properties are conventional in the sense that there are alternatives that are observationally equivalent. Consider the topological difference between a cylinder and an open, flat plane. Life on the cylinder would be repetitious if you circled around in the direction perpendicular to the axis. You would keep returning to the same place. So does repetition of this kind serve as evidence of a topologically closed space, closed in at least this dimension? No, because the evidence is equally consistent with life on a flat, open space that has repeating bands of identical properties. What seems like returning to the same place is really just coming to the identical point in the next band. Because all the appearances would be identical between the two different topological spaces, this property of open or closed is underdetermined by the evidence.

In the case of topological conventionality, all of the cases of ambiguous evidence are hypothetical cases. There are not any real observations on the books that are topologically ambiguous. There is no evidence of repetition that forces us to make a choice between a cylinder and a banded plane. The conventionality of topology is an abstract concern about what we would have to say, and the element of choice we would have, if such evidence did appear.

Furthermore, many topological features are not conventional. They are not underdetermined by the evidence. Whether the space is continuous or discrete is such a property, and it is this topological property that is responsible for the conventionality of geometry. The four-dimensional spacetime of events is continuous. There are no gaps in physical space or in time, no atoms of space or of time. Spacetime is infinitely divisible. So between any two points in spacetime, say, between an event at one end of a train car and an event at the other, there is an infinite number of other spacetime points, an infinite number of places that something could be. There is the same number of spacetime points, namely an infinite number, between an event at one end of the train car and an event at the midpoint of the car, or between here and the sun. We cannot measure distances by counting the intervening spacetime points, because every interval, no matter how long or short, has the same number of spacetime points, an infinite number. Spacetime itself, because it is continuous, offers no intrinsic way to determine measurements of distance, and that is why it has no intrinsic geometry. If spacetime were discrete and grainy, we could determine distance simply by counting the grains between here and there. We would not have to impose any conventions about measurements because spacetime itself would determine the measurements. But it is not discrete, and it is not grainy.

What does the conventionality in the theory of relativity mean about our ability to know appearance or reality? The principles of conventionality were implications from the theory itself, and so they are exactly the sort of epistemological issues we are concerned with. Does the unavoidable freedom to choose such important things as the one-way speed of light and the nature of measuring sticks introduce a pervasive subjectivity in the resulting description of nature? It does not, at least not in a way that the description is no longer about the way nature is.

An analogy to our use of language shows that conventionality does not preclude the possibility of describing the facts of nature. A language is a tool of description and we are free to choose which of the many languages to use. We can make the choice on the basis of which language is more useful and more appropriate to the situation at hand, or we can make the choice to suit our whims. This element of choice is analogous to the choice available in geometry. And the conventionality in both cases shows an aspect of what is right about the claim that our descriptions of nature are always in part subjective. It is up to us whether we say water freezes at 0°C or at 32°F. But of course once we choose one temperature scale or another, in a sense, one language or another, we can no longer choose what to say in that language. We cannot choose to say that water freezes at 1000°F or at 0°F. Once the initial choice of language is made, then nature dictates what is true and what is false. We choose *how* to say it, but we cannot choose *what* to say.

In a similar way, it is up to us whether we say spacetime is curved or spacetime is flat. The freedom of choice is of the nature of measurement, and once we have adopted one measurement language or another, then we are forced to describe nature in certain ways. If we choose to say that there are rigid measuring tools and that light travels along geodesics, then we must say that spacetime is curved. Once we have adopted the language of unchanging measuring sticks, then the description of nature is dictated by nature itself. This shows that there is a limit to the human influence on the description of nature. We cannot say just anything we want, since the description must be consistent under the constraint of nature. There are two contributions to the description, our own and that of nature, and we can tell which is which. It was the physics of spacetime that recognized the human contribution, the element of choice. The physics itself contributed to our being able to distinguish those aspects of our description of nature that depend on us from those that depend on the way nature is.

SUMMARY

On the issue of relativity and realism, we can summarize both what relativity has to say about epistemology and the limits of knowledge, and what epistemology has to say about relativity. On the first question, there is nothing in the physics of relativity to support Bohr's claim that physics in general does not or cannot describe nature as it really is. The relativistic model of the world does not imply that our knowledge is limited to appearance. On the second question, the theory of relativity turns out to be a clear example of the possibilities of objective knowledge. The special theory of relativity illustrates the necessity of observing the world from a reference frame and the freedom in choosing whatever reference frame we want. It also articulates the effects of one's reference frame, one's perspective, on the values of measurements. But once a reference frame is chosen, properties have determinate, objective values and the values of measurements are set by the nature of things. The special theory of relativity thus shows the human component of the description of nature, and the natural component. Our analysis of the general theory of relativity does the same. The theory serves to point out the element of conventionality so that we know what part of the final description is of the natural world.

Chapter 6
Quantum Mechanics

—————————————— • ——————————————

Quantum physics leads to the rejection of determinism.

Alastair Rae,
Quantum Physics: illusion or reality?

For [quantum mechanics], the truth is you can only talk about the universe in terms of probabilities. You can never, by definition, be certain about it.

James Burke
"The New Physics" in the BBC series,
The Day the Universe Changed

The first of these comments on quantum mechanics is somewhat misleading. If it means that with quantum mechanics all natural processes are non-deterministic, then it is simply false, and we will soon see why. The second remark is not even false; it is a contradiction in itself. One thing is clear though; the interesting thing about quantum mechanics, and what makes it a challenge to our usual understanding of knowledge and reality, has to do with probability and indeterminism. We will start then with some basic ideas on probability and the concept of determinism to get these straight before they become complicated by quantum mechanics.

PROBABILITY, CAUSE AND EFFECT, AND DETERMINISM

Probability and causation are aspects of the natural world that are quite common in our daily lives. The probability of rain this afternoon is seventy percent. Smoking may cause lung cancer, and it certainly causes resentment among the non-smokers in the restaurant. We are familiar with the basics of probability and the concept of cause and effect. Nonetheless, there may be some lurking confusion and some important questions yet to be asked and answered. We need to get those out in the open before turning on the quantum mechanics.

A good way to talk about probability is to talk about playing with dice, fair dice, dice that are not loaded or chipped. A die is fair if the outcome of rolling it is entirely random. The number 1 turns up just as often as 2 or 3 or any of the other numbers on the cube. Each of the six possible outcomes is equally likely on any one throw. Expressed as a probability in which zero corresponds to impossible and one to dead certain, the probability that the one-dot side will show up on any one toss is 1/6. We can abbreviate this to save time and space.

$$P(1) = 1/6$$

The probability that the five-dot side will show up is also 1/6.

$$P(5) = 1/6$$

The probability that *either* the one or the five will turn up is 1/3. This is because, of all the six possible outcomes, each of which is equally likely, two of the outcomes satisfy the specified condition. If a one shows up, we win. If a five shows up, we win. So our chances of winning are 2 out of 6, and 2/6 equals 1/3.

$$P(1 \text{ or } 5) = 1/3$$

Now suppose we roll the die twice, noting both the first and second showings. There are 36 possible outcomes, since, for each of the six possible outcomes of the first roll, there are six possible outcomes of the second. Thus, the probability of getting a 2 on the first roll and a 3 on the second is 1/36.

$$P(\text{first 2, then 3}) = 1/36$$

What is the probability that the *sum* of the first and second outcomes will be 2? There is only one way this could happen, if the first roll is a 1 and the second roll is a 1. So the probability is 1/36.

$$P(\text{sum} = 2) = 1/36$$

There is no way to get a sum less than 2. So,

$$P(\text{sum} = 1) = 0.$$

What is the probability that the sum will be 6? There are five ways to do this: first a 1 and then a 5, first a 2 then a 4, 3 then 3, 4 then 2, and 5 then 1.

$$P(\text{sum} = 6) = 5/36$$

We can also ask about the *conditional* probability. If the first roll has already been done and it turned out to be a 1, then what is the probability that the sum of the first and second rolls will be a 2? In other words, on the condition of the first roll being 1, what is the probability the sum will be 2? The answer is 1/6, since, of the six possible outcomes of the second toss, only one, getting a 1, will make the sum equal to 2. In the abbreviated notation, the information on which the outcome is conditioned follows a vertical bar. Read it as, the probability that the sum is 2, given that the first roll is a 1, is 1/6.

$$P(\text{sum} = 2 \mid \text{first} = 1) = 1/6$$

As a pop quiz you should figure out for yourself why

$$P(\text{sum} = 6 \mid \text{first} = 6) = 0$$

and

$$P(\text{sum is greater than 3} \mid \text{first} = 3) = 1.$$

This ends the preliminary exercise in using probabilities, and we can now go on to questions about probabilities in general. For example, how is probability measured? We can talk about the probability on a single throw, the probability, for example, that a 1 will show up. But the actual outcome of a single throw is either a 1 or a 2 or any of the other numbers. There is nothing one-sixth about the outcome. If a 1 shows up we can say the probability of a 1 was 1/6, and if a 1 does not show up we can say the probability of a 1 was 1/6. So how can we check to see if the theorized value of 1/6 is right?

The only way to do this is to make many tosses and see that the outcomes are random, that is, that each possible outcome occurs with the same frequency. Similarly, the only way to measure the probability of a particular outcome is to toss the die many times and see what fraction of all results is the outcome you are interested in. Roll the die many times and you will see that a 1 comes up one-sixth of the time. Probability is measured as the relative frequency of outcomes of many trials.

If probability can be measured only as a relative frequency of many events, then perhaps that is all probability really is. We could think of probability as a property only of large groups of events, many tosses of the die. On this interpretation, a single event does not have any probability about it. Describing the probability of a single toss of the die is only loose talk, shorthand for the relative frequency of outcomes if we were to do many tosses. This suggested account of what probability is, is motivated by a desire to keep our conclusions about the way things are as close to the evidence as possible. If all of the evidence about probability is from measurements on large ensembles of events, then what reason, what evidence, is there to claim that there is such a thing as the probability of a single event?

The only way to make sense of probability as a property of a single event is to say that it is a hidden property, one that we cannot directly measure. If a fair die has a determinate propensity, a disposition, to turn up with a one-dot showing, with an inherent chance of $1/6$, this is a property we cannot directly observe. We always observe individual outcomes to be definitively one-dot, that is, seemingly $P(1) = 1$, or not one-dot, $P(1) = 0$. Any interpretation of probability as being a property of individual events entails a significant gap between reality and appearance. The probability *is* a feature of each individual event, though no individual toss of the die *appears* probabilistic. This is important. If we maintain that probability is property of each individual roll of the die, then we participate in a description of nature, an account of how things really are, that goes beyond how things appear.

So far we have suggested two possible interpretations of the nature of probability. One attributes it only to ensembles of events, the other to single events. They have in common the idea that probability is an objective feature of the world and not a feature of our attempt to know about the world. That is, both objective interpretations of probability put it in the things themselves, either as a group or as individuals, whether or not someone is observing or trying to know which face of the die will turn up. There is a third interpretation of probability that regards it as a subjective feature of our knowledge. Probability measures our lack of knowledge of the situation. Saying that the probability of something is $1/3$ is not talking about nature, because the outcome is entirely determined. Either it will happen or it will not; we just do not know. Given the information and understanding we have of the situation, it seems to us one-third likely that it will happen. If we knew everything, all probabilities would be zero (for things that will not happen) or one (for those that will).

Here is an example to motivate this subjective interpretation of probability. You have arranged to meet your friend Bob for lunch in a restaurant at noon. Walking towards the appointed place, you wonder, what is the probability that Bob is at this moment already in the restaurant? There is a fact of the matter here. Bob either is in the restaurant or he is not in the restaurant, you just do not know. There is nothing chancy or indeterminate about the situation itself. So when you say to yourself, the probability is $1/10$, you are measuring the state of your information of the situation. You know the time (it is exactly noon) and you know Bob (he is notoriously tardy and irresponsible), so you figure there is little chance he is on time. More information, like a look inside the restaurant, would change the probability without changing the object (Bob) itself. Less information, if you did not really know Bob and you forgot the time you were supposed to meet, would push the probability toward the maximally uncertain fifty-fifty, P(Bob is inside) = 1/2. Clearly, the probability is a measure of your uncertainty here, and not a propensity of the object itself. It is a subjective probability.

Perhaps the case of rolling the dice is the same as meeting Bob for lunch. Perhaps, if you knew *everything* about the die and the toss, then you would know with certainty how it was going to turn out. You would have to know the exact shape of the die, the way it was held, the forces on it from the hand and gravity, the air currents, and so on. But then its trajectory would be as uniquely determined as that of a tossed tennis ball. The point of landing of the ball is fully determined by the initial conditions, angle of ascent, force, and the like. And there seems to be no good reason to suppose that the die is any different. The details of its landing, namely which face is up, are determined by the initial physical conditions. The only uncertainty is in our knowledge of the situation. If we knew everything about nature, there would be no probabilities. This is where Einstein's famous admonition, "God does not play dice with the universe," figures in. He does know everything (God, not Einstein), and omniscience, on this subjective interpretation of probability, eliminates the need for probabilities.

The three accounts of the nature of probability thus fall into two distinct groups. There is at this time no need to decide which account is true. All that is important is to be clear on the distinctions so that when we encounter probabilities in quantum mechanics we can pause and ask which notion of probability is being discussed. Objective probability is a property of nature itself and applies to events that are genuinely not deterministic. It is an aspect of metaphysics since it purports to describe the way things are. Subjective probability is a property of our knowledge of nature and applies to cases where we lack information and so we are uncertain. It is an aspect of epistemology since it purports to describe the way we know about things.

We can put this distinction in Bohr's terms. Objective probability describes how nature is. Subjective probability describes only what we can say about nature.

We can also put this discussion in terms of cause and effect. The outcomes of events that we have been interested in, the die coming up 1, or Bob being in the restaurant, are the effects. Among the initial conditions, the shape of the die, the force, and so on, is the cause. It is customary to divide the initial conditions into those factors that are merely the background conditions, and that one (or perhaps those few) that are actually the cause. If a house has burned down, for example, many factors were involved. There had to be oxygen to feed the flames, and flammable fuel to do the same. These are just the background conditions and as houses stand day-in and day-out, *not* burning, these conditions are present.

Clearly these conditions are necessary for a conflagration, but not sufficient. You need a spark or a match to initiate the flame. That is the cause
of it all. But of course the spark by itself is not sufficient either. You need
the fuel and the oxygen. Thus, of all of these initial conditions, each is
necessary but none is sufficient, and so it is not entirely clear how to mark
one off as *the* cause.

Perhaps the distinction between cause and conditions is just a matter
of our own interests and pragmatic concerns. The fire marshal, interested
in the basic, physical events, might simply cite the match as the cause.
The insurance company, interested in the homeowner's responsibility,
might single out the pile of oily rags in the garage. The district attorney,
interested in the possible culpability of an arsonist, will focus on that person's actions, the striking of the match, as the cause. And so on. Singling
out *the* cause is not a straightforward matter. Luckily, this aspect of ambiguity is not a pressing issue in quantum mechanics.

More important in quantum mechanics and its relevance to appearance
and reality is the issue of determinism, whether the initial conditions, regardless of which condition we identify as the cause, uniquely determine
the outcome. As with realism, it is wise to distinguish the *question* of determinism from determinism as a way things are, that is, determinism as
one possible *answer* to the question. The question of determinism is this:
Given the state of affairs in nature at one moment (the initial conditions),
is there a single, unique state of affairs that can happen in the next moment? The key is the uniqueness. A plurality of possibilities is a case of
non-determinism.

The "yes" answer to the question of determinism, the claim that nature
is at each juncture deterministic, is what Einstein had in mind when he
spoke of God's aversion to gambling. Each causal beginning determines
a unique effect. If the particular cause happens, then the particular effect
has to happen, period. Therefore, any talk of probabilities must be of subjective probabilities, because in nature everything is exactly predeter-
mined from the initial conditions. There are no such things as objective
probabilities.

The "no" answer to the question of determinism, the claim that nature
is, at least in some cases, non-deterministic, is where Alastair Rae and
many others say we will be led by quantum mechanics. The rejection of
determinism entails that in some natural cases, a causal setup can have
more than one possible effect. Which one of the possible effects happens
is a matter of irreducible chance that we can describe only in terms of
probabilities. In this case the probabilities are objective, features of the
cause and effect relation itself. It is not that we do not know which outcome will happen; it is that nature itself has an element of chance.

Again, we should not be trying to align ourselves with one of these
ways of looking at nature or the other. That would introduce a bias into

the evaluation of quantum mechanics that is to come. We might find in quantum mechanics whatever appeals to our prefigured favorite view of the world. We do not want that to happen. At this point it is only the clear distinction that we want to bear in mind, the distinction between deterministic and non-deterministic processes. It is also important to realize that determinism is a question or a claim about dynamic processes, about changes from one moment to the next. It makes no sense to ask whether the world is deterministic at an instant.

PARTICLES AND WAVES

There are two ways that energy and information can be sent from one place to another. Some energetic signals in nature are in the form of waves. Others are in the form of particles where physical bits of stuff actually make the trip from the one place to the other. A signal that goes as a wave goes without any object traveling from the source to the reception. A wave is a pattern of disturbance in the intervening medium. If you and I hold opposite ends of a taut string, I can send you messages and enough energy to ring bells or to get you to do things, simply by moving my end of the string up and down. No bit of string goes from me to you. In fact, none even moves in your direction. They move up and down without moving along the horizontal line between us. By doing this they pass along the disturbance I have created as a shape in the string. The moving shape is the wave. The string is the medium. Waves can take on all sorts of shapes. I can send you a single, sharp pulse, a slow, broad pulse, a series of pulses at a certain frequency, and so on.

Waves and particles have some irreconcilable differences. That is, waves routinely behave in some ways that particles simply can never behave. Likewise, particles have some properties that are fundamentally contrary to the nature of waves. It is therefore impossible to form a hybrid of the two, a signal that is part wave and part particle. There is no middle ground. A signal must be either a wave or a particle.

As one of the irreconcilable differences, note that waves exhibit the behavior known as interference. Particles do not. Interference is easiest to understand by considering two sources of waves that converge on a single spot. This is two kids in a bathtub and making waves in the water, or two pieces of string knotted together onto the end of a third, all forming the shape of a Y. On this latter arrangement I can send you signals from both my left hand and my right, though waves from either source end up on the single third string in your hand.

Imagine that I am sending you periodic signals with both hands, and my hands are moving up and down at the same rate and in sync. When my left hand goes up, so does my right. When the left goes down, so does the right, and over and over. In this case, the two wave sources have the same frequency and are in phase. The waves they create have equal wavelengths, the distance between adjacent peaks. When the two waves come together at the knot, their peaks will coincide and their valleys will coincide, as shown in Figure 6.1. The result on the third string will be doubly-high peaks and doubly-deep valleys. In other words, the two waves come together in phase and reinforce each other. This is constructive interference and you get a doubly intense dose of energy at your end.

Constructive interference is not so unlike what you would expect of particles. Send two and you double the result. But destructive interference is just the opposite and it could never happen with particles. Send two and the result is zero. With waves this will happen when the two waves some together exactly out of phase. If my right hand is going up as my left hand goes down, and then my right goes down as my left goes up, the two sources are exactly out of phase. When the two waves arrive at the knot, one pulls up as the other pulls down. They cancel each other out and the knot moves neither up nor down. The knot and the rope you are holding carry no energy, no wave. As shown in Figure 6.2, two waves have come together and disappeared. Particles never do this.

There is another way to create destructive interference in the waves. If the two sources are in phase but one is further away from the intersection by a distance of half a wavelength, then the two waves will be out of phase when they meet. If my two hands move up and down in sync, but one of my strings is longer than the other by half of a wavelength, then my two waves cancel at the knot and you get nothing. If my two strings differ in length by exactly one whole wavelength, then the two waves will arrive at the knot in phase and interfere constructively. You get a doubly strong signal.

The bottom line on interference is this: When waves come together the result depends on their relative phase, whether they are pulling together or in opposition. Particles do not have phases, and so they can never interfere destructively. They never come together and cancel. If two bullets

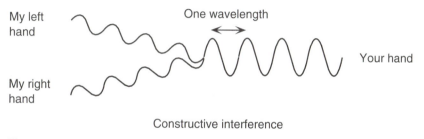

Constructive interference

Figure 6.1

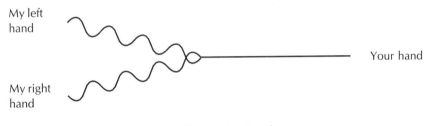

My left
hand

My right
hand

Your hand

Destructive interference

Figure 6.2

are speeding your way, there is no chance they will come together and disappear. Your toast is up.

Now the question is, which of these two kinds of signals is light? It is certainly a process of sending energy and information from one place to another, and from the special theory of relativity we know how fast the information moves. But other than speed, the special theory of relativity pays little attention to the nature of light. The theory is amenable to transmission either as waves or as particles of light. It is true that we presumed that light and electro-magnetic radiation in general behave as waves, and this was the dominant belief at the time of the development of the special theory of relativity. But it could also be done with a model of light as particles, with tiny bits of energy pouring out of your flashlight with a speed of 3×10^8 m/s, like molecules of water from a firehose.

If the special theory of relativity is neutral with respect to light being a wave or a particle, how can we know which of these modes of transmission light really is? This is not a matter of taste or fashion of aesthetic predilection. It is a matter of experiment and coherent theory. One experiment in particular, well known at the inception of the special theory of relativity, clearly shows light behaving as a wave. It shows that light can interfere destructively, with two beams coming together and canceling out. Two beams of light intersect and there is darkness. It is called the double-slit experiment, and is easy enough to set up and get unambiguous results, so easy that it is routinely done by first-year physics students in a two-hour lab session.

The "double-slit" refers to a mask that is placed in front of a source of light, a mask in which two tiny slits have been cut. The setup is shown in Figure 6.3. The two slits are there simply to give us two sources of light that are close together and, if light is a wave, in phase. We know they would be in phase because they are pieces of the same initial wave and they are equidistant from the original source. The two sources of light shine onto a screen some distance away. In the physics student's lab, the screen is usually about 1 meter from the double-slit mask, and the two slits are about one-tenth of a millimeter apart.

Before turning on the light we should ask what is to be expected if light is in fact a wave, and what is to be expected if light is a particle. If the prediction of what will appear on the screen is different for the two models, then this will be a crucial experiment to decide which model is accurate.

If light is a wave, then the two sources are analogous to my two hands making waves on two strings. The knot is at the screen. At the center of the screen, the point opposite the mid-point between the two slits, the two sources of light are equally far away. We expect a very bright spot, because the two waves will arrive in phase and constructively interfere. For points on the screen that are above the midpoint, the lower slit will be further away than the upper slit. There will be a point at which the lower slit is further away by a distance of half a wavelength of the light. Here we expect the two waves to arrive out of phase and to interfere destructively. Here we expect darkness. There will be a dark spot on the screen an equal distance below the center point too, where the upper slit is further away by half a wavelength. Moving farther down the screen, the upper slit is getting even further away. At the point on the screen

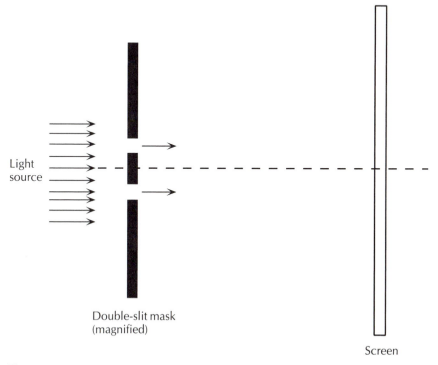

Light
source

Double-slit mask
(magnified)

Screen

Figure 6.3

where the upper slit is further away than the lower slit by a distance of exactly one wavelength, the two waves arrive in phase and we expect another bright spot of light. Of course, there will be one at an equal distance above the center of the screen as well. Above that, where the lower slit is further away by one and a half wavelengths there will be another dark spot. And so on.

In sum, here is what to expect if light is a wave. Because of the phenomenon of interference, there will be a series of bright spots and dark spots, the center being bright. This is shown in Figure 6.4a.

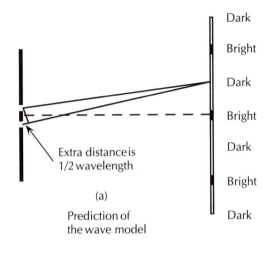

(a)

Prediction of
the wave model

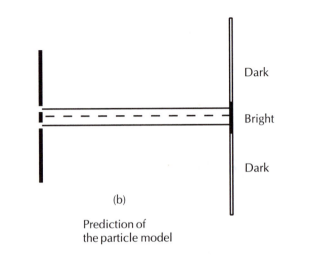

(b)

Prediction of
the particle model

Figure 6.4

Now suppose that light behaves as a particle and predict what the screen will look like when you turn things on. There will be no interference, so wherever the light from the two sources comes together there will be bright light on the screen. The light will spread out a bit from each of the slits, since it will be spreading out from the original source, so the glow from each slit will be somewhat blurred and spread out on the screen. The results of the two slits shining together will just be the sum of their individual images. There is no phase relation to worry about. Assuming that the blur of the two slits overlap, we expect one big bright spot at the center of the screen. There is no repetition of bright and dark. This prediction is shown in Figure 6.4b.

When you do the experiment, when you turn on the light, the results are those predicted by the wave model, every time. There is a clear series of bright spots and dark spots. The double-slit experiment is unambiguous; light behaves as a wave and not as a particle, at least in these circumstances.

If physicists had stopped here and left well enough alone, the description of nature would be just fine, neat, and orderly. We could say that light is a wave and bullets are particles. But it is not in the nature of scientists to leave well enough alone, and this is to their credit. The key to being objective, the key to being scientific is to be always willing to do another experiment even when issues seem to be settled. Another experiment with light, the so-called photo-electric effect, was one of the eye-openers that initiated quantum mechanics. While relativity began largely with conceptual and logical concerns of theory, quantum mechanics came to life with the spark of experiment.

The photo-electric effect in general is the phenomenon in which light shining on a metal causes electrons to fly off the metal. It is the primary step in a photo-voltaic cell or the electric eye that senses your coming and opens the door to the grocery store. These things work by converting light energy into an electric current, that is, by the photo-electric effect.

The photo-electric effect works because metals have a lot of loose electrons. These electrically charged particles are not tightly bound to any individual atoms and are relatively free to move around the metal. This is also why metals conduct electricity so well. Shining light on a metal gives these free electrons energy and makes them move around on the metal. If an electron gets enough energy it can move fast enough to fly off. This is the photo-electric effect.

Again we should ask what to expect of the photo-electric effect if light is a wave. We will be looking for correlations between the properties of light (things like how bright it is and what color it is) and properties of the electrons that come off (how many there are and how fast they are going). If light is a wave, then the energy it delivers is proportional to the intensity, the brightness of the light. We would expect that the intensity of the light would therefore be correlated to both the number of electrons liberated, and to their speed. Brighter light will free more electrons and they will be moving faster.

This is not at all what happens when the light is turned on in the photo-electric effect. The speed of the electrons has no correlation to the intensity of the light. Instead, the speed of the electrons is affected by the frequency of the light, that is, by its color. It does not matter if the light is bright or dim; what matters is its color. Brighter light releases more electrons, but not speedier ones. As the frequency increases and the color changes from red to violet, the energy of the electrons increases and so their speed increases.

The most perplexing aspect of the photo-electric effect is the existence of a threshold frequency. For each particular photo-electric setup, that is, for each particular metal used, there is a frequency of the incident light below which the effect simply does not happen. No matter how bright the light, and no matter how long it shines on the metal, if its frequency is below the threshold, not a single electron comes off. This is contrary to what is expected of waves. Since waves deliver energy continuously they have a cumulative effect. As time goes on, the energy should build up and release a few electrons. But this does not happen. If the frequency of light is steadily increased, there is no effect until the threshold is reached, and then suddenly they pour off. The flow of electrons jumps from zero to millions. As the frequency of light is increased further, the rate of electrons stays the same but their energy, measured in terms of their speed, steadily increases.

The results of the photo-electric effect are shown in Figure 6.5. Predictions on the belief that light is a wave are shown with dashed lines. The actual experimental results are shown in solid lines.

The photo-electric effect thus confounds the theory that light is a wave. Furthermore, on a suggestion of Einstein's, we can make good sense of the photo-electric effect if we regard the light as hitting the metal in discrete bundles, that is, as particles. What we measure as the frequency of light is proportional to the energy carried by each particle. Brighter light, higher intensity, means more particles but not higher energy. This explains why shining brighter light on the metal does not increase the energy of the electrons, because each is liberated by interaction with a single particle of light and gets the amount of energy set by the frequency of that single particle.

Particles of light also account for the otherwise curious threshold frequency. The electrons are held onto the metal with an amount of energy distinctive to the kind of metal. Without this binding energy the electrons would just drift off spontaneously. The threshold frequency corresponds to the energy of an individual light particle that just equals this binding energy. Light particles with lower frequency have insufficient energy to

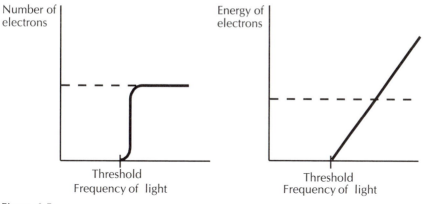

Figure 6.5

break the bond. They cannot liberate any electrons. Electrons freed by light particles with frequencies higher than the threshold are sent on their way with whatever energy is left over after overcoming the binding energy.

In short, light does not behave as a wave in the photo-electric effect; it behaves as a particle.

The theorized particles of light are called light quanta or photons. "Quanta" is plural for "quantum" which means the smallest possible amount, the unit of a discrete quantity. A penny is a quantum of money in the United States, since there is no smaller amount that one can have or exchange. To make the most we can of the word, we can say that money is a quantized commodity. It is not continuous; it comes in discrete units, quanta.

Einstein's suggestion then is that light is a quantized commodity. Each quantum of light, each photon, carries a tiny amount of energy (they are more like Italian Lire than U.S. pennies). Normal sources of light like a flashlight or a star are producing billions and billions of photons in a steady stream. Because there are so many and each is energetically so small, light seems continuous to us, just as the stream of individual molecules of water feels continuous when you wash your hands.

So is light a wave or a particle? The double-slit experiment shows it to be a wave. The photo-electric effect shows it to be a particle. No single measurement shows it to be both, and none could, since waves and particles have mutually exclusive properties. Under some circumstances light behaves as a wave. Under other circumstances light behaves as a particle. Under no circumstances does it behave as both. The property of being wave-like, in other words, like the property of being particle-like, is relative to the circumstances of measurement. Light is a wave with respect to interaction with a double-slit apparatus. Light is a particle with respect to interaction with electrons in a metal.

We are now doing quantum mechanics, and just as in relativity, the conceptual key is learning how best to ask questions of nature. The moral of the story of the special theory of relativity is that a question like "What is the length of this thing?" is incomplete. Length is relative to a reference frame, so the question must be, What is the length with respect to such-and-such reference frame? Length is a two-place property, a relative property, and so the question of length, as well as the specification of length, must include both pieces of information. Now, in quantum mechanics the emerging point is that the question, "Is light a wave?" or "Is light a particle?," is incomplete. Being wave-like and being particle-like are two-place properties, relative properties. The way to ask the question in order to get an informative answer has to be, Is light a wave with respect to such-and-such measurement? As with reference frames, we have complete freedom in choosing which experiment to do. That much is entirely up to us. But once that choice is made, the results are determined by nature itself. There is a determinate fact of the matter. If you elect to do a double-slit experiment, the outcome will be a wave. It is like litmus paper. The color cannot be red and blue at the same time, but the paper has the propensity to turn either red or blue, depending on the nature of the liquid we put it in. We decide the interaction to reveal its color, but then there is a natural fact as to what color will emerge.

There is no denying that this is surprising and counterintuitive news about light and the property of being wave-like or particle-like. By common sense these ought to be intrinsic properties such that something either is a wave or not, once and for all. But by the same common sense, length ought to be an intrinsic property too. Quantum mechanics, it turns out, will challenge common sense in the same way that relativity did.

The complicated nature of light can be summarized: The description of light must be in terms of incompatible properties that cannot be observed together at the same time but that together give a complete model of the behavior of light. The properties of being a wave and being a particle are mutually incompatible, yet the complete description of the nature of light must include both. The two properties complement each other.

There is another important concept that emerges from these experiments, the concept of properties that are not simultaneously measurable. If we set up an interaction to measure the wave aspects of light, we lose information on its particle-like properties. There is no way, in principle, to measure both at once. It is not our fault. It is not a fault of imprecise machines or insufficient powers of perception. It is in the nature of things. It is the nature of being a wave that precludes showing signs of being a particle.

One might wonder whether this peculiar complementary nature is unique to light. Recall that light has a unique role in the special theory of relativity. It alone has a speed that is not relative to reference frames. And so it is not unreasonable to expect light to have a distinctive status in quantum mechanics. It is not unreasonable, but it turns out to be wrong.

Louis de Broglie speculated in 1923 that if light, a phenomenon we had thought to be a wave, sometimes acts like a particle, then perhaps things we take to be particles, bits of matter like atoms and electrons, sometimes act like waves. This would be idle speculation and not a matter of credible science if it were not testable. The best way to test wave-like properties of something is with a double-slit experiment. This will reveal the destructive interference that is characteristic of waves if anything will.

The dimensions of the double-slit experimental setup are important. The separation between the two slits must be on the order of one wavelength of the waves passing through. It can be twice as far apart as one wavelength, or maybe even ten times as far, but when it gets to be a hundred or a thousand wavelengths between the two slits, the pattern of bright and dark spots on the screen is so compressed that the brights cannot be distinguished from the darks. The slits must be roughly the right distance apart to make the interference pattern visible.

The details of de Broglie's suggestion of wave behavior of matter indicate that the wavelengths would be very small, on the order of the size of an atom. To test this, a double-slit mask of this spacing is required, a spacing too small to construct. Luckily, there are such masks ready-made in nature. A crystal of something like salt or ice is an array of regularly spaced atoms. Thinking of the atoms themselves as the opaque parts of the mask, and the spaces between the atoms as the slits, a line of three atoms in a crystal acts as a tiny double-slit mask. The distance between slits is the size of the atom, that is, roughly the wavelength of de Broglie's predicted electron waves.

The experiment to test de Broglie's suggestion is shown in Figure 6.6. Electrons are sent into the crystal just as light was shined onto the double-slit mask. The focus of attention is then on the screen. One big pile of electrons in the center would indicate that electrons behave as particles. A series of small accumulations and empty spots, the analog to bright and dark spots, would indicate interference among the electrons and hence that they are behaving as waves.

The results indicate waves. An unambiguous interference pattern shows up on the screen, forcing the conclusion that in these circumstances, electrons behave as waves. The experiment has been done on other objects besides electrons, but always very small things like protons or alpha particles (the nucleus of a helium atom). The results are always the same. In a double-slit type interaction, these objects behave as waves. The reason the experiment is always done with very small things is that the wave-

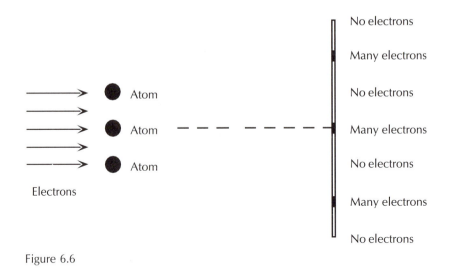

Figure 6.6

length of waves associated with pieces of matter gets smaller as the mass of the object gets larger. So for macroscopic things like a tennis ball or your brother, things you might like to double-slit experiment on yourself, the required separation between the slits is an unattainably small distance.

Since de Broglie's suggestion of objects acting like waves in some circumstances has turned out to be true for anything we are able to test, the concept of wave/particle duality must be generalized to describe everything, not just light. Everything, all matter and transmissions of energy like light, must be described using mutually exclusive, complementary properties. The idea of incompatible properties that are not simultaneously observable is universal. Light is not unique in this way. Light is not special in quantum mechanics.

THE STATE FUNCTION, COMPLEMENTARITY, AND THE UNCERTAINTY PRINCIPLE

It is one thing to say that bits of matter behave under some circumstances like waves; it is another thing to fill in the details of this model. For example, what is the frequency of an electron wave? What is the wavelength? In general, what is waving? With waves on a string it is easy to make sense of such properties as the amplitude, the frequency, and the wavelength of the wave. The amplitude is the displacement of the string from its normal straight shape. If I move my hand a large distance we get a large amplitude. Frequency is the rate I move my hand back-and-forth and wavelength is the distance between humps in the string. We could take a snapshot of the wave on the string, put a ruler on it and measure the physical distance that is the wavelength.

It is almost as easy to make sense of these properties of light waves. The amplitude of the wave is the strength of the electric field, caused by moving a charged particle, and subsequently the cause of moving other charged particles. The frequency is the rate the source charge is moving back and forth. And the wavelength is the distance in space between adjacent points of strongest electric field pointing in the same direction.

It is not at all easy to make sense of these properties for the waves associated with matter. Precise measurements of the spacing of bright and dark spots on the screen of a double-slit experiment can be used to measure the wavelength of waves passing through the slits. By this means, together with measurements of the momentum of the electrons entering the crystal, it is discovered that the wavelength and the momentum are inversely proportional. The bigger the momentum, the smaller the wavelength, and the ratio between the two is the same in every case, no matter what the object is. The constant of proportionality is called Planck's constant, after Max Planck, and given the abbreviated symbol h. Planck's constant is unimaginably small, $h = 6.6 \times 10^{-34}$ Joules-seconds. The units, the Joule-seconds, are not so important to us. The smallness though, the fact that this key number in quantum mechanics has the decimal point, then 33 zeros, then the 66, shows that the quantum effects we are talking about are apparent only in the realm of extremely small objects.

Just for the record, we can put this relation between the wave-like property of wavelength and the particle-like property of momentum into the shorthand of an equation. Wavelength is traditionally abbreviated with the Greek letter λ, and momentum by a lower-case p.

$$\lambda = h/p$$

This gives us a physical interpretation of the wavelength of the waves associated with bits of matter, but we still do not know about the amplitude. A wave is a series of peaks and valleys, a moving shape, so what are these matter waves peaks and valleys of? What physical quantity is strong where there is a peak and weak (or perhaps strong but in the opposite direction) where there is a valley?

The first thing to do is to give the amplitude of the wave a name and then to figure out what in the world it corresponds to. It is called the state function and is represented with the Greek letter Ψ. Different values of Ψ at different points in space make up the shape of the wave. So if Ψ is zero at one point and then gets steadily larger at points to the right, reaches a maximum and shrinks to zero, then becomes a negative number, hits bottom and returns to zero, the state function is the S-shaped wave shown in Figure 6.7.

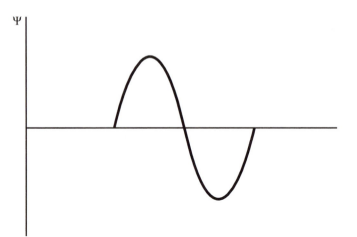

Figure 6.7

Any shape is possible, and different physical circumstances, depending on the mass of the object and the forces acting on it, result in different shapes of the state function, different distributions of the value of Ψ in space. The shape of the state function changes over time as the object itself changes its speed or position or whatever. There is a certain state function that corresponds to an electron orbiting the nucleus of an atom. Another one for a proton stuck in the nucleus itself. And so on. The business of quantum mechanics is in figuring out the shape of the state function in various situations and to see how it changes over time. Everything there is to know about an object is in the shape of the state function.

But what physical property does Ψ represent? What exactly is it about the object that has no value at points where Ψ is zero and is strong where the shape of the state function takes on large values of Ψ? Ψ itself has no direct correspondence to any real physical property, but the squared value Ψ^2 does. The value of Ψ^2 at each point turns out to be related to the probability that the object is at that point. It can be related to dynamic probabilities of properties other than position as well.

Figure 6.8 shows a possible state function distributed along one horizontal spatial dimension. It is important to know that the displays of Ψ and Ψ^2 are both abstract mathematical plots. These are graphs of the values of Ψ and Ψ^2, not snapshots of actual shapes of things in the physical world. Nothing takes on a wavy shape; rather the changing values of Ψ vary from point to point, from large to small, to zero to negative to zero, and so on. Furthermore, the plots of Ψ and Ψ^2 catch them at just one instant in their dynamic evolution. At another instant the shapes will be different. The evolution of the state function is generally smooth and deterministic. Except in special circumstances, there are no abrupt jumps from one shape to another, and the shape of the wave at one instant uniquely determines the shape it will have in the next. In this sense, quantum mechanics is a deterministic theory. In this sense, quantum physics does not force the rejection of determinism.

Notice in Figure 6.8 that squaring the values of Ψ keeps all the zero values at zero and converts all the negative values to positive. The physical interpretation of the Ψ^2 graph is this. If you were to look for the object at point A at this instant, there is no chance of finding it. If you were to look instead at point B, your chance of finding the object is pretty good. B is the most likely place it will be. In mathematical terms of probability, P(object at A) = 0. And since Ψ^2 is greater at B than at C, P(object at B) > P(object at C).

The fact that the state function of the object is spread out in space does not necessarily mean that the object, the electron or whatever, itself is spread out in space. Every observation of the object shows it to be entirely at one point in space or another. Observing an electron is like rolling dice. The probability of rolling a one-dot face is 1/6, but that does not

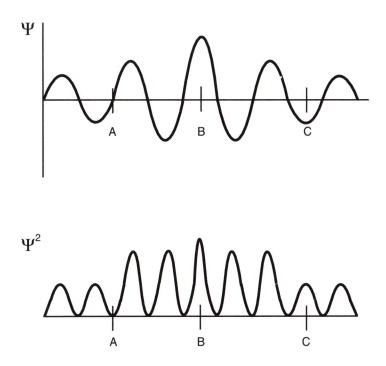

Figure 6.8

mean that you ever get one-sixth of a one-dot face. You always get all of one face or all of another, just as when you observe an electron it is always entirely at one place or another. Actually rolling the die, like observing the electron, resolves the ambiguity of the probability. Once the observation is made, there is no more probability because the electron is where you observed it to be. The spread-out wave function has collapsed to a single point, the point where the electron was observed. It could have collapsed to a different point, just as the rolled die could have come up with a different face showing. With fair dice, each face is equally likely, but with electrons and other quantum mechanical things, some places are more likely than others. The state function keeps track of these relative probabilities.

It is the square of the state function that is correlated to a physical property, the probability of being at one place or another, but it is the state function itself that determines the interference effects when objects collide. When waves come together as in a double-slit experiment with electrons in a crystal, the result is the sum of the two state functions. Figure 6.9 shows a case of two state functions Ψ_1 and Ψ_2 that are exactly out of phase and of the same amplitude over the considerable length from point A to point B. In that interval the two waves interfere destructively and the result is a state function of value zero over that space. Even though the individual waves give a probability of finding an electron in the interval, together they cancel the probability to zero. This would be a dark spot on the screen.

Even considering a single object, that is, a single state function with no interference effects, the question of where the object is has become complicated. The position of an object has a single determinate value only in

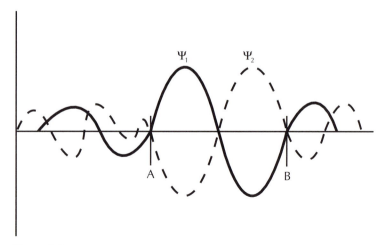

Figure 6.9

the special case in which the state function has the degenerate shape of Figure 6.10a. In this case, the probability that it is at the point x is one, and the probability that it is at any other point is zero. This is the shape of the state function after position has been measured to be at the point x. For this particular state function there is no uncertainty or indeterminateness in its position. It is at point x, and we know it is at point x.

It will be helpful to keep track of this quantity, the indeterminateness of position. Using the letter x to be the *value* of position, we can use a Greek delta to be the indeterminateness, the spread of the value. In this case then, $\Delta x = 0$.

What is the momentum of the object whose state function is shown in Figure 6.10a? Recall that the correlation between the shape of the state function and the momentum of the object is the wavelength. Shorter wavelength means greater momentum. So what is the wavelength of the state function in Figure 6.10a? It is neither long nor short; it is nonexistent. The wavelength, and hence the momentum, is completely indeterminate.

What sort of state function would represent an object with an exact value of momentum? For momentum to be without ambiguity, the wavelength of the state function must be a single, constant value. That is, the length between each pair of adjacent peaks must be the same so there is no ambiguity in the value. The state function must be the shape shown in Figure 6.10b. There is no indeterminateness of momentum in this wave, and recalling that momentum is abbreviated with the letter p, $\Delta p = 0$. The position of this object though is completely indeterminate. The wave pattern extends forever along the x-axis in order to determine the constancy of the wavelength, thus the object could be found at any point where Ψ is not zero. There is an infinite number of places where the object is equally likely to be. Information on where it is, or even where it is likely to be, has been entirely lost. Δx in this case has become infinitely large.

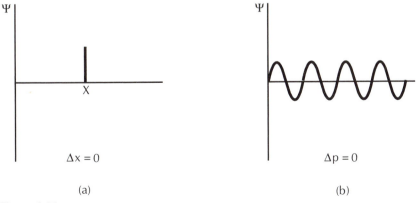

Figure 6.10

These two cases, Figures 6.10a and 6.10b, together indicate that there is a trade-off in determining position and momentum. If one is exact, the other cannot be. An object cannot have an exact position and an exact momentum at the same time. This trade-off is related to the irreconcilable differences between being a wave and being a particle. Momentum, since it is directly related to wavelength, is a wave-like property. Position, since it is a localized, point-like quality, is a particle-like property. So the incompatibility between position and momentum is a manifestation of the incompatibility between the particle and wave nature of things. This sort of complementarity is a key component of the quantum mechanical account of nature. The complete description of something, whether it is light or a bit of matter like an electron, must be in terms of incompatible properties that cannot exist at the same time. We knew from classical theory and observations that wave and particle are mutually incompatible properties, but from quantum mechanics we realize that no description of an entity is complete without both. And we knew from classical theory and observation that a description of an entity requires both position and momentum. Now we have learned from quantum mechanics that these two are mutually incompatible. There will be other pairs of incompatible properties in quantum mechanics, and many manifestations of this principle of complementarity. The description of nature must include incompatible properties that cannot exist at the same time but that together give a complete model.

Just as wave properties and particle properties are not simultaneously observable, momentum and position of an object are not simultaneously measurable, at least not with exact precision. The two cases shown in Figure 6.10 are the extremes in which Δx is zero in one case, or Δp is zero in the other. For cases in between, Δx and Δp are both finite. As one gets larger, the other gets smaller. This is the trade-off, and it is expressed by the inequality known as the **Heisenberg Uncertainty Principle**.

$$\Delta x \cdot \Delta p > h/2\pi$$

The h is Planck's constant, again showing that these effects show up only in the realm of the very small. The 2π is there because of a general feature of waves. The uncertainty principle says that the product of the inexactness of position and momentum (not of the values of these properties, but of the size of the spread of values) must be greater than the number $h/2\pi$. If one property becomes more exact, the other must become less exact. Δx and Δp cannot both be zero at the same time, for then their product would be zero in violation of the uncertainty principle.

We arrived at the uncertainty principle through an analysis of the way things are, not the way things are measured. The trade-off between determinate position and determinate momentum is a property of waves, not a property of how we observe waves. The inequality that is the Heisenberg uncertainty principle holds whether or not anyone is measuring position and momentum. Thus, the name *uncertainty* principle is somewhat misleading. Uncertainty implies an inexactness in our knowledge of things, as I am uncertain whether or not Bob is in the restaurant. Uncertainty is an epistemological fuzziness, but the inexactness in position and momentum is deeper than that. It is metaphysical. The object cannot *have* an exact position and an exact momentum at the same time. One or the other *is* indeterminate, not just uncertain to us. A more accurate name, more accurate than uncertainty principle, would be the principle of indeterminateness.

SPIN AND THE EPR EXPERIMENT

Light waves, like waves on a string, can have the property of being polarized. While holding my end of the taut string, if I move my hand consistently up and down, the wave pattern is always straight up and down, never left and right, never at any angle other than straight up and down. The waves are vertically polarized. If I move my hand back and forth in the horizontal plane, I make waves that are horizontally polarized. The string moves in the horizontal plane. Light waves can be polarized by moving the charged particle that creates them consistently back and forth along the same line. Have the charge move up and down and the electro-magnetic wave is vertically polarized. And so on.

There is also a kind of polarization associated with particles like electrons. Quantum mechanical particles have a property called spin, because they behave as if they are spinning like tops. It is important to keep in mind that it is only a loose analogy between quantum mechanical spin and macroscopic spin as we see in tops or the earth on its axis. It is only *as if* the elementary particles are spinning, and in some important respects quantum mechanical spin is quite different from the spinning we are used to.

One thing they have in common is that spin, both macroscopic and quantum mechanical, gives the object a distinctive direction. Since the earth spins on its axis it has a particular orientation with respect to the sun. The axis of spin is useful for identifying this orientation. A spinning top is the same way. When it is standing vertical on the table, its axis of spin is vertical. When it falls over, its axis is horizontal. It is customary to identify the orientation of the spin with the orientation of the axis. So, the standing top is spin-up, and the fallen-over top is spin-horizontal. And if the top could flip entirely over so that its north pole was directly below its south pole, it would be spin-down.

Elementary particles like electrons and protons have an axis like a spinning top that allows us to identify the direction of their orientation. To say an electron is spin-up means it behaves as if it is spinning around an axis with its north pole up. Every electron has the same magnitude of intrinsic spin. It is as if they are all spinning at the same rate. But their spin-orientations, the direction the axis is pointing, can differ.

Spin orientation is easy to measure. Just as the spin of the earth gives it a magnetic field, the spin of an elementary particle gives it a magnetic field. Recall that a moving electric charge creates a magnetic field. The movement of spinning of the electric charge that is the electron creates a magnetic field. Each electron is like a tiny bar magnet with a north pole and a south pole, and the orientation of the magnet is the orientation of the spin. Thus, even if the object is too small to see, as is the case with an electron, its orientation can be measured magnetically.

There is an important difference between the bar magnet and the electron, a difference that shows up when we measure their orientations. The bar magnet can have any number of possible orientations; an electron is restricted to only two. If we had a random collection of bar magnets and sent them through a device that sorted them according to their initial orientation, we would get a full spectrum of values. Specify a coordinate system, and some of the magnets will be spin-up, some spin-one-degree-from-up, some spin-two-degrees-from-up, ... , some spin-down, and so on. No orientation is left out. The electrons behave much differently. If we send a random collection of electrons through a device for sorting them according to spin orientation, there are only two possible outcomes. Each electron comes out either as spin-up or as spin-down, regardless of what direction we specified for our coordinate system. No electron comes out at any orientation in between. This is shown in Figure 6.11.

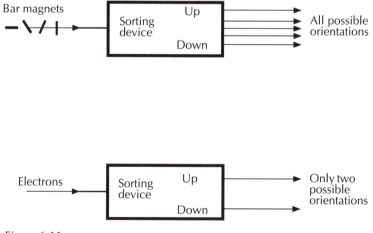

Figure 6.11

Because there are two possible orientations, the electron spin orientation is called a two-level system and the determination of which of the two possible orientations has occurred is a two-level measurement. The device for performing such a measurement of spin orientation is a Stern-Gerlach device. Its principle working parts are a magnetic field, usually situated with the field lines oriented in the vertical plane, and two detectors that count the number of electrons. Send a bunch of electrons into the magnetic field and some will behave as if they are spin-up, and the rest will behave as if they are spin-down. These are the only two choices. The characteristics of the magnetic field in a Stern-Gerlach device are such that the spin-up particles are physically separated from the spin-down particles, and each group is sent to its own detector. In this way the device measures the relative frequency of spin-up particles and spin-down particles.

The quantum mechanical state function has the information on the probability of getting spin-up or spin-down results in a Stern-Gerlach measurement. Considering the initial conditions of the electrons entering the device, how they got there, and what has been done to them recently, quantum mechanics can perfectly predict the relative number of spin-ups and spin-downs. But for a single electron entering the device, quantum mechanics cannot in general predict whether it will be spin-up or spin-down. Like a single roll of the die, we can only give the probability of the outcome.

This raises the question about quantum mechanics that is at the heart of the issue of appearance and reality. What is the nature of the probability that is inherent in the quantum mechanical description of the world? Is it an objective probability reporting on irreducible chanciness in the world itself, a God who plays dice? Or is it a subjective probability, a reflection of insufficient information on our part, a weakness of knowledge and an indication that more work needs to be done? Answering this question will get us back to Bohr's distinction and his claim about the proper task of physics. It will get us back to appearance and reality. It is so important that one more macroscopic analogy is in order to make sure the choices are clear.

Imagine an election for President of the United States in which there are only two choices on the ballot, a Democrat and a Republican. With no third party or independent candidates on the ballot, and no write-in votes allowed, the act of voting is a measurement with two possible outcomes, Democrat (D) or Republican (R). In this sense it is a two-level measurement. A third choice on the ballot would make voting a three-level measurement.

With this in mind, imagine a room full of people who have not yet voted but who are dedicated party-line voters. To make the example less abstract, say there are twenty people in the room, ten Democrats and ten Republicans. The voters are anonymous and unrecognizable in the sense that nothing about their appearance indicates their political persuasion. Only by their voting can we tell whether they are Republican or Democrat.

When an individual emerges from the room and steps up to vote, we know before the act that the probability of voting Republican is 1/2. After seeing the result of the vote (We can record each vote as it happens rather than wait for the totals on the nightly news), the probability of voting Republican has changed to 1 (if the individual did vote R) or to 0 (if the vote was D). The measurement of voting has not changed the individual in any way; it has only changed our knowledge of the individual. The property that is each individual's political affiliation is determinate both before and after they vote.

The room full of Democrats and Republicans can be called, naturally enough, a mixture. In the language of quantum mechanics it is a mixed state.

Now imagine a different situation in which the room is filled with non-partisan voters. Given the constraints of the ballot, each of these people will have to vote either R or D, but each is at this point undecided. Each is a kind of superposition of political opinions, a superposition of Democrat and Republican. Since these people are all alike in this way, the group can be called a pure state.

Assume that each non-partisan voter is as likely to go Democrat as Republican when he or she comes to the snap decision in the booth. As an individual emerges from the room and heads for the voting area, we know that the probability of voting Republican is 1/2. But once the vote has been cast and recorded, the probability collapses to 0 (if the vote was for the Democrat), or to 1 (if the vote was for the Republican). Our knowledge and our description of the situation is no different in this pure-state case than it was for the mixed state.

There is, though, a significant difference in what really happens to the individual (and the group) in the process of voting. The non-partisan has been changed and can be clearly identified after the vote as either a Democrat or a Republican. Their indecisive superposition of voting choices has been focused into one or the other choice. As the pure-state room full of non-partisans files out, votes, and is regrouped in another room, it has changed into a mixed state, a room full of decidedly Republican voters and decidedly Democratic voters.

The important thing to notice about this election is that no measurement of the voters, neither individual votes nor statistical results of the group, can distinguish the pure state from the mixture.

Measuring spin orientation is in some important ways like measuring the votes of these people. Just as there is no way to tell of a particular person whether she will vote Republican or Democrat, there is no way to know before the measurement whether a particular particle will be spin-up or spin-down. In the case of quantum mechanical spin, though, the uncertainty is not simply imposed on the example to fit heuristic purposes. The uncertainty is a central feature of the quantum theory. The theory is definitive in that the measurement of spin orientation will show one of the two possible values, up or down, and the theory can tell us the probability for each result. Before the measurement, however, the theory cannot identify the ups from the downs. This is one of the mystifying features of quantum mechanics. Where other scientific theories make definitive predictions about the properties of things, quantum mechanics offers only probabilities.

The inability to describe and predict pre-measurement values of spin orientation (among other properties) is a sign of weakness to some commentators on the theory. Einstein, for example, thought it showed that the quantum theory was overlooking some features of the particles that would reveal their spin orientation before measurement. There must be some "hidden variables," some property of the particle that determines its spin orientation before entering the Stern-Gerlach device but that quantum mechanics does not reveal. He claimed the theory was incomplete in this way, and, with Podolsky and Rosen, proposed a thought experiment to prove the incompleteness. The famous Einstein-Podolsky-Rosen (EPR) experiment was given a simplified version by David Bohm, a version with no loss of implication for the nature of quantum mechanics.

In Bohm's version of EPR, two spin-orientable particles are produced from a single event so that the spin orientations of the two are correlated. The source of the two particles has no spin at all. It is, for example, a molecule composed of components whose collective spin orientations cancel each other out. The total spin is always conserved, so the net spin of the entire system must always be zero. If one of the produced particles measures spin-up, in other words, the other must cancel that by being spin-down. Quantum mechanics is clear to say that total spin is conserved.

Such a process is easily realizable in nature, and, in fact, this version of EPR has been done. Bohm suggests that the molecule (net spin = 0) could disintegrate into two pieces, each with non-zero spin. The combined spin of the subsequent two-particle system will still be zero if the two are pointing in opposite directions. Conserving spin demands that if one is up the other is down.

The EPR experiment continues by allowing the two particles to separate by a significant distance, as shown in Figure 6.12. The quantum mechanical description of the situation is that at this point the system of the two particles is a pure state. The two particles are identical in that each is a superposition of up and down. It is not that one is up and the other down and we just do not know which is which; rather, they are both non-partisan at this stage.

Particle 2

Particle 1

Figure 6.12

There is more to come in the EPR experiment. Now measure the spin orientation of one of the particles. Call it particle 1. Before this measurement, all the quantum theory could do for us was report a probability of 1/2 that particle 1 would be spin-up. P(1 is spin-up) = 1/2. After the measurement though, we know exactly what the spin-orientation of particle 1 is. Furthermore, and this is the exciting part, we also know the spin-orientation of particle 2 even though it was never measured. If particle 1 measures up, then particle 2 must be down. But if both particles were in the indeterminate, neither-up-nor-down state until particle 1 was measured, what happened to particle 2 to make it spin-down? There is no problem understanding the physical act of measurement on particle 1 changing particle 1 from a non-partisan to a determinate spin-upper. (Well, there is, but it is another problem for another chapter. It is called, appropriately enough, the measurement problem.) The problem at hand (sometimes called the locality problem) is understanding how the measurement on particle 1 caused particle 2, instantly and at a significant distance, to change from a non-partisan to a determinate spin-downer.

We are left with a choice between two explanations. We could accept these, as Einstein called them, "spooky actions at a distance," as part of the natural world, or we could say that particle 2 was always a spin-down particle (and particle 1 was always spin-up) and the measurement on particle 1 changed our knowledge of both particles but did not change any physical property of either. Einstein advocated the second choice and concluded that quantum mechanics was incomplete. The two-particle system was a mixed state all along, but quantum mechanics was unaware of the variables, the hidden variables, that marked particle 1 as spin up and particle 2 as spin-down even before either was measured.

The choice is either a pure state of particles with indeterminate spin orientation, or a mixed state of particles with determinate spin orientations that we just don't know. It is a contrast between spin orientation being *indeterminate* before the measurement, and spin orientation being determinate but *uncertain*. It is non-partisans or inscrutable party loyalists. The ambiguity in pre-measurement spin orientation is either in the physical nature of things or in our imprecise knowledge of things.

In modern terms, the EPR experiment proves that either there are hidden variables in nature (and quantum mechanics is an incomplete theory because it cannot describe them), or interactions in nature can have a feature of non-locality in the sense that what happens to one object can influence another, instantaneously and at a considerable distance.

Einstein, common sense, and classical (pre-quantum mechanics) physics all favor the hidden-variables choice. Under this description, the two particles would have determinate states of spin orientation all along. The state of particle 2 and the outcome of a measurement of its spin orientation would be independent of the measurement on particle 1. There would be no interaction at all between 1 and 2, once they are separated. If this is true, and quantum mechanics is incomplete, the theory is nonetheless entirely successful in predicting the outcomes of measurements on large ensembles of events. It is a statistical theory with a perfect record of success. It is still the theory that is the basis of everything you would call high technology. To say it is incomplete is not to fault its considerable pragmatic success. It is rather an issue of metaphysics, that the theory has not gotten to the bottom of things.

Incompleteness of the theory results only on the hidden-variable choice of the either-or results of the EPR experiment. There is still the choice of non-locality to consider.

BELL'S PROOF

Since the result of the EPR experiment is a disjunction, an either-or choice between hidden variables and non-locality, it is no surprise that it raised more disagreements than it settled. Bohr disagreed with Einstein as to the right choice in the EPR results. Bohr argued that quantum mechanics is complete, that there are no hidden variables lurking outside the theory that determine the spin orientations of the two particles. Though it seems curious to us, Bohr claimed, each particle is in a superposition of spin-up and spin-down states before being measured. They are non-partisans rather than inscrutable party loyalists. Indeterminateness and non-locality are simply aspects of the quantum world we have to live with, according to Bohr.

The debates over the proper interpretation of the EPR experiment, whether to side with Einstein or with Bohr, continued without much progress on either side until the 1960s and Irish physicist John Bell. Bell believed that Einstein was right and he hoped to clarify and support the hidden-variable conclusion of the EPR experiment. He ended up doing just the opposite. In 1964, with what is known as Bell's theorem, he proved conclusively that locality and hidden variables cannot both be true of nature. His results, based on a thought experiment that is a variation on the thought experiment of Einstein, Podolsky, and Rosen, ruled clearly against Einstein's claim that there are hidden variables and that interactions must all be local. We are stuck with spooky actions at a distance and with inherent indeterminateness of properties at the level of quantum systems.

In the original EPR experiment, there is no measurement of spin orientations we can do that can distinguish whether the two particles are in a pure state before measurement, or in a mixture of states. That is, there is no way to tell if they are identical and each a superposition of spin-up and spin-down, or they are different in that one is spin-up and the other spin down. It is the same inability we had with the voters, since the outcome of voting was the same for the group of non-partisans as it was for the mixture of Democrats and Republicans. It is this inability of the experiment that results in the ambiguity and the choice of how to interpret the outcome.

Bell's accomplishment begins by expanding on the original experiment in a way that can distinguish pure states from mixtures. In the original experiment, only the vertical, up-and-down, orientation of spin was measured or theorized. We measured the vertical orientation of particle 1 and from this knew the vertical orientation of particle 2. If #1 measured spin-up, then #2 must be spin-down. When the vertical spin-orientation of particle 2 is then measured, it does in fact always turn out to be opposite of particle 1. We always talk about spin-up and spin-down, but there is nothing special about the vertical direction. The Stern-Gerlach device itself can be rotated to any orientation. If the magnetic field inside the device is horizontal rather than vertical, the EPR experiment is the same only in terms of the particles being spin-right or spin-left. Any orientation will do, but in the original setup it is always the same orientation for both particles. If #1 is measured in the vertical orientation, so is #2. If #1 is measured at 45°, so is #2.

As long as all of the spin orientation measurements are done with the Stern-Gerlach device at the same orientation, there will be no way to tell whether the pair of particles is a mixture or a pure state. Bell's suggestion was to do the experiment with measurements of spin orientation at a variety of angles, sometimes with the Stern-Gerlach device to measure #1 at a different orientation from the Stern-Gerlach device to measure #2. With this setup we will be able to ask questions like, If particle 1 is measured to be spin-up, what is the probability that particle 2 will be spin-right? By rotating the Stern-Gerlach devices, we will also be able to measure the spin orientation of particle 1 at a variety of angles, not just up-and-down. Of course, for each individual particle we can measure spin orientation only once, since measuring is a kind of interaction that affects the correlation between the two particles. Each measurement, in other words, records the information only by altering the specimen so that it no longer has the information. It is like an archaeological dig, where to understand the site you must take it apart. The point is that to measure the spin orientation of particle 1 at various angles we will need a whole series of particle 1's. This experiment will require many trials with identically prepared particles, and the results will be about the relative frequencies, the statistics, of spin orientations.

With this expanded setup of the EPR experiment, measurements will be able to distinguish whether the pair of particles is a pure state or a mixture. That is, we will be able to tell whether or not each particle has a determinate spin orientation before the measurement. We will be able to decide between Einstein's local hidden variables and Bohr's indeterminateness and spooky actions at a distance.

There are lots of different orientations of the Stern-Gerlach devices that work to prove Bell's point. The version of Bell's experiment presented here is taken from David Mermin's exposition of Bell's theorem. As one example, let us do the experiment such that the devices have three possible orientations. They can be oriented on the vertical axis (as in the original EPR experiment), or on an axis rotated by 120° from the vertical, or on an axis rotated by 120° in the other direction from the vertical axis. Figure 6.13 shows these three possible orientations as seen from the approaching particle. The view you see in Figure 6.13 is the view from the cockpit of particle 1 or particle 2. The axis that is straight up-and-down is marked with a V, for vertical. The spin-up end is marked with a plus sign (+). The axis that is rotated in the clockwise direction, that is, around to the right, is marked with an R. Its spin-up end, the end of the Stern-

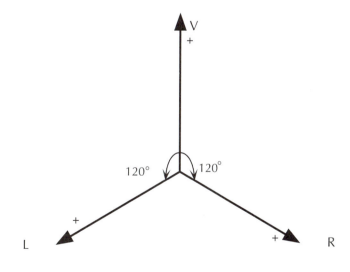

Figure 6.13

Gerlach device that has the north pole of the magnet, is similarly marked with a plus sign. The axis that is rotated in the counter-clockwise direction, that is, around to the left, is marked with an L, and its north-pole, spin-up end has a plus sign.

An electron flying towards its measurement at the Stern-Gerlach device will be flying straight into the view in Figure 6.13. It will encounter a magnetic field that is either aligned with the V-axis, or aligned with the R-axis, or aligned with the L-axis. Our freedom to measure at any one of these three orientations is what distinguishes Bell's setup from the original EPR experiment.

As in the original EPR setup, the two particles are produced from a source that has net spin of zero. Therefore, the spin orientations of particle 1 and particle 2 must always cancel out. If #1 is spin-up along the R-axis, then #2 must be spin-down along the R-axis. Having the Stern-Gerlach devices at opposite ends set along the same axis, both along R in this example, is just doing the old EPR experiment but with the whole setup tilted a bit. As before, there will always be perfect canceling of the two spin orientations when measured along the same axis. Things get interesting, and we get new information about nature, when the two Stern-Gerlach devices are oriented along different axes.

Assume, for the sake of argument, that there are local hidden variables that determine the spin orientations of the particles at all times in a way that requires no signaling, no connection of any kind between events at opposite Stern-Gerlach devices. The individual particle might encounter any one of the three orientations when it is eventually measured, so it must have a determinate property of up (+) or down (−) along each of the three axes. As particle 1 is heading towards its measurement device, there must be a determinate factor of the particle as to how it would align with a V-oriented field (up or down), how it would align with an R-oriented field, *and* how it would align with a L-oriented field. If there are the sorts of hidden variables Einstein believes, the sort that prevent non-locality, then no possible spin orientation can be indeterminate.

As an example, still assuming that there are local hidden variables and all values of spin orientation are determinate, suppose particle 1 has the property that it will be spin-up if measured in the V-axis. It must also have determinate properties as to how it will be (up or down) if measured along either of the other axes. Let's say it is spin-up along L and spin-down along R. Among the properties of particle 1 are its speed, its mass, its electric charge, and the hidden variable of its spin orientations at various angles. We can abbreviate for this specific example by saying that particle 1 is (V+ L+ R−), meaning that it is preset to be spin-up along V, spin-up along L, and spin-down along R. The theory of local hidden variables requires that each of these three slots be filled in with a determinate up or down, because anyone of these three measurements might be done. And of course, if particle 1 has the property of being (V+ L+ R−), then particle 2 must have the property of being (V− L− R+). This is necessary to maintain the zero net-spin of the whole system, the two particles together.

On the other hand, if there are no local hidden variables, the spin orientation of either individual particle is indeterminate until measured. And once measured, only one of the orientations is determinate. Before encountering the Stern-Gerlach device, particle 1 has the property of being a superposition of up and down, for each of the three axes. If we measure along the L axis and it turns out to be spin-up, it is still a superposition of up and down for the other two axis. Only one of the three slots can have a determinate values at any time, and that only after measurement. Of course, if particle 1 is measured to be L+, then particle 2 must be L−. Particle 2 has no determinate spin orientation along V or R.

With these details of the two theories in place, we can do Bell's suggested experiment. The procedure is to do many repetitions of the experiment with different settings of the two Stern-Gerlach devices at the particle 1 end and the particle 2 end. Not only do we keep changing the orientation of each device after each trial, but we make sure they are never both at the same orientation for the same trial. If #1 measures spin orientation along R, then #2 must be set to measure spin orientation along V or L. Over the many trials we will keep track of how often particle 1 and particle 2 show the same spin orientation along their respective axes, that is, they are both spin-up or both spin-down, and how often the two particles differ in their spin orientation, one up the other down.

The two theories, hidden variables and indeterminate superpositions, predict different outcomes. The predictions are in the form of probabilities, namely the probability that the spin orientations will be the same, and the probability that the spin orientations will be different. These probabilities can then be measured as relative frequencies to see who is right.

To see what is predicted by assuming that the particles have no determinate spin orientations until measured, it is best to start with one special case and then generalize to all cases. As the special case suppose the Stern-Gerlach device to measure particle 1 is set along the V-axis, and the device for particle 2 is set along R. Furthermore, suppose #1 is measured to be spin-down. That is, particle 1 is V−. We know that particle 2 would be V+ if it were measured along V, but it is measured along R. What is the probability that particle 2 is R+? In other words, what is the probability that the two particles will have different spin orientations, #1 being down and #2 being up, when they are measured along these different axis? This is a conditional probability we are looking for:

$$P(\#2 \text{ is } R+ \mid \#1 \text{ is } V-) = ?$$

Focus on particle 2, because we already know all we need to know about particle 1, namely that it is V−.

The one thing that is known about particle 2, that it is V+, is not the one thing we want to know, whether it is R+. So the task is to figure out the relation between spin orientation in one direction (V) and the probability of spin orientation in another direction (R). In general this relation can be quite complicated, but for the particular case we have set up, with the angle between the two orientations set at 120°, it is not so bad. We can figure out what the relation is, and then find the conditional probability above, by first considering the easiest possible cases.

The question is, if the particle is spin-up along the vertical axis, then what is the probability it will be spin-up along another axis that is rotated by some angle Θ from the vertical? We will work our way up to Θ being 120°. If $\Theta = 0$, then the new orientation we are talking about is itself vertical and the question is, If the particle is spin-up along the vertical axis, what is the probability that it is spin-up along the vertical axis? This is not a trick question; it is my promise to start with easy cases. The answer is 1, it will certainly be spin-up.

This first case is shown in Figure 6.14a. In all diagrams in Figure 6.14, the bold arrow shows what we know to be true of the particle, namely that it is spin-up along the vertical axis, and the thin line represents the axis of spin orientation we want to know. In Figure 6.14a, $\Theta = 0$ and the two lines are aligned.

The next case, with $\Theta = 90°$, is shown in Figure 6.14b. The vertical spin-up of the particle is halfway between the up (+) and the down (−) ends of the horizontal axis. Like a coin balanced on its thin edge, this one could go either way. There is nothing in the vertical spin-up property to make a measurement along the horizontal axis more likely to come out one way rather than the other. Thus, the probability that it will be spin-up along this horizontal direction (spin-left in figure 6.14b) is 1/2.

If $\Theta = 180°$, the probability is zero. This is shown in Figure 6.14c. The question in this case is, If the particle is spin-up in the vertical direction, what is the probability it is spin-up when measured by a machine that is rotated top to bottom? It amounts to asking, If it is spin-up along a vertical axis, what is the probability it is spin-down along a vertical axis? Again, it's no trick. The answer is 0.

These are three easy cases to show that the probability of the particle being spin-up along an axis that is an angle Θ from the vertical is a function of Θ. In general the probability depends on the cosine of the angle Θ :

$$P(\text{spin-up along an axis at angle } \Theta) = [\text{cosine} (1/2 \; \Theta)]^2$$

All we need this formula for is to figure out the probability for the particular case of $\Theta = 120°$, the case of our setup of the Bell experiment. This

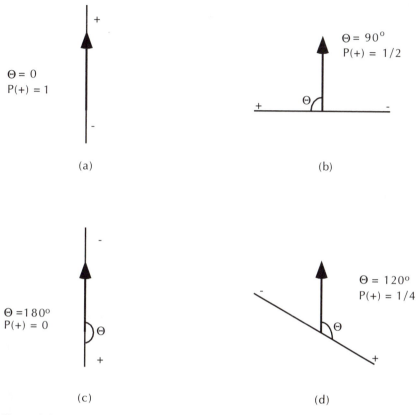

Figure 6.14

is shown in Figure 6.14d. One-half of 120° is 60°, and the cosine of 60° is 1/2. The square of this is 1/4. This is the conditional probability we need. The R-axis is rotated 120° from the V-axis. We know that particle 2 is V+ because particle 1 has already been measured to be V−. So, if particle 1 is V−, the probability that particle 2 is R+ is 1/4.

$$P(\#2 \text{ is } R+ \mid \#1 \text{ is } V-) = 1/4.$$

This result can be generalized to all cases of measuring the spin orientation of particle 1 along one axis and the spin orientation of particle 2 along a different axis. In every case, the probability of getting a different value for the two particles, that is, one of them spin-up along its axis and the other spin-down along its axis, is 1/4. This is because each of the three axes, V, R, and L, are separated from the other two by 120°. The calculation we did for the one case holds for any case because of the symmetric way we have set up the Bell experiment.

So here is the important result, the payoff from the calculation of conditional probabilities. On the assumption that there are no local hidden variables that determine the spin orientations of the two particles at all times, the probability that the spin orientation of the two will be different (one up, the other down) along different axes is 1/4.

P(spin orientation is different for #1 and #2) = 1/4

This is the prediction of quantum mechanics itself, since it does not recognize any local hidden variables that determine spin orientations. Remember that all quantum mechanics can tell us about the spin orientations of particles 1 and 2 before they are measured is the probabilities of outcomes. The calculations of probabilities in the absence of factors that really predetermine the spin orientation is exactly the quantum mechanical description of the situation.

This is also the way things turn out when the experiment is done. Measured in terms of relative frequency, it turns out that one-quarter of the time, the spin orientations of particle 1 and 2 come out to be different, one up and the other down. This is interesting information, but it really was not necessary to actually do the experiment. Bell's point can be made just as well with a thought experiment, one based on logic and the credibility of other ideas, as was the original EPR experiment. As we will see, the assumption of local hidden variables leads to probabilities that are different from those derived by using quantum mechanics. In other words, quantum mechanics cannot be incomplete because it lacks hidden variables. It cannot be augmented, because it is incompatible with local hidden-variable theories. Because of the different probabilities predicted, any theory that includes local hidden variables is incompatible with quantum mechanics and with the flawless record of empirical success of that theory. Quantum mechanics works. Its predictions of probabilities have always been right. Its empirical success in all these other experiments should be enough to rule out alternatives like the local hidden-variable theories that are empirically incompatible. If the assumption of local hidden variables leads to different predictions, the assumption is very likely to be wrong.

Quantum mechanics predicts that particle 1 and particle 2 will have different spin orientations along different axes one-fourth of the time. What does a local hidden-variable theory predict?

The key difference between quantum mechanics with indeterminate spin orientations and a local hidden-variable theory is that in the latter, every particle always has a value of spin orientation along every axis. It is important to point out that we are working with only the most general features of a local hidden variable theory. *Any* theory that describes particles 1 and 2 as having in themselves the determinate information on spin orientation and as having this information all the time and without the influence of any distant, instantaneous signals from the other, will be included in this analysis. Locality (no instantaneous signals from a distance) and hidden variables in general are on the line here.

Particle 1 has a locally determinate spin orientation. So if we measure particle 1 to be spin-down along the vertical axis, that is, it is V−, it must also have a value (not a probability) of spin orientation along R and L. It might be, for example, (V− L− R−). Or it might be (V− L− R+). There are other configurations as well. What is important is that each particle has some definitive value of spin orientation in each of the three slots. In this type of theory the separate particles have their determinate properties all on their own. They do not rely on each other for their spin orientation. Their state-functions are separately complete, in isolation. This is what we mean by a *local* hidden variable.

If particle 1 has the property (V− L− R−) then particle 2 has the property (V+ L+ R+). This is one possible pair of particles. Call it **pair A**. The other possible pair mentioned above is particle 1 with the property (V− L− R+) and so particle 2 is (V+ L+ R−). Call this **pair B**. Again, there are lots of possible pairs, but we shall follow just these two through the experiment.

pair A   **pair B**
particle 1: (V− L− R−)   particle 1: (V− L− R+)
particle 2: (V+ L+ R+)   particle 2: (V+ L+ R−)

Suppose that **pair A** has just been created at the center of the experiment and each particle is heading toward its Stern-Gerlach device to be measured for spin orientation. We know that they will be measured along different axes, since their respective Stern-Gerlach devices must be oriented along different axes, but we do not know exactly which one is on which axis. We only know they are different. What is the probability that one of them will be spin-up (+) and the other spin-down (−)? This is easy. Particle 1 will be spin-down on any of the three axes, and particle 2 will be spin-up on any of the axes. So regardless of the setting, they will *always* have different results. The probability of being different is 1. The details of these outcomes are shown in Table 6.1. There are six possible ways to have the two Stern-Gerlach devices set up. The one for particle 1 could be along the L-axis and the one for particle 2 could be along the V-axis. This possibility is the first entry on the table. If this is the setting, #1 measures spin-down (−) and #2 measures spin-up (+), as shown on the first line of the table. The other five possible settings and their results fill out the table. As promised, all settings of the Stern-Gerlach devices result in #1 being spin-down and #2 being spin-up. For **pair A**,

Table 6.1

Predicted Outcomes of Bell's Experiment Using a Local Hidden Variable Theory

Stern-Gerlach settings		Outcome pair A		Outcome pair B	
#1	#2	#1	#2	#1	#2
L	V	−	+	−	+
L	R	−	+	−	−
V	L	−	+	−	+
V	R	−	+	−	−
R	L	−	+	+	+
R	V	−	+	+	+
		6 different		2 different	
		P(different) = 1		P(different) = 1/3	

P(spin orientation is different for #1 and #2) = 1

But this is just for **pair A**. What about other possible pairs of particles, for example, **pair B**? The table also shows the outcomes of the six possible settings for **pair B**.

Of the six settings of the Stern-Gerlach devices, four come out with the same spin orientation for particles 1 and 2, and two come out different. Thus, the probability of getting different spin orientations is one-third. For **pair B**,

P(spin orientation is different of #1 and #2) = 1/3

Every possible pair of particles will turn out either like **pair A** or **pair B**. To prove this, all you need to do is figure the outcomes for all the possible pairs as we did for **A** and **B**. They will all have different results for spin orientation of #1 and #2 either all of the time or 1/3 of the time. In general then, a theory of determinate spin orientation values predicts different results in *at least* 1/3 of the trials.

P(spin orientation is different for #1 and #2) ≥ 1/3

This is the definitive result. Quantum mechanics, with the claim of indeterminate spin orientations, predicts different values of particles 1 and

2 will show up in one-fourth of the trials. Any local hidden variable theory, with determinate spin orientations, predicts this will happen no less than one-third of the time. One-fourth is less than one-third. Thus, with this experimental setup, Bell has proven that no theory with local hidden variables, no account of nature that says that an object has determinate spin orientation before we measure it, no such theory is compatible with quantum mechanics. Furthermore, experimental results agree with quantum mechanics. Local hidden variables lose.

The details of Bell's proof are perhaps somewhat daunting. It is not so important to have followed all the calculations of probabilities, as long as you believe the results and follow the logic.

To summarize Bell's proof, recall that the original EPR experiment left us with a choice. Either the spin orientations of both particles 1 and 2 are determinate in each particle from the start (and the information on what those orientations are is hidden from the quantum mechanical perspective), or the spin orientation of #2 becomes determinate under the influence of some instantaneous (non-local) interaction from the measurement on #1. Local hidden variables or non-locality. Bell's extended experiment shows that the first choice is impossible. Bell proves that it is impossible to have both determinate spin orientation (hidden variables) and locality.

Bell's proof rules out any theory of local hidden variables, but we should note it does not rule out a theory of *non*-local hidden variables. We will encounter David Bohm's version of a non-local theory in the next chapter. For now, suffice it to say that a non-local hidden-variable description of particles 1 and 2 would say that there is a determinate fact of the matter of spin orientation for both particles. It is fully deterministic as to how each will emerge, up or down, from their respective Stern-Gerlach devices, but the definitive information for each is not fully intrinsic to each in isolation. These hidden variables are *contextual*. The spin orientation of particle 1 will depend on the orientation of the Stern-Gerlach device it encounters. This much contextuality is entirely local and is not at all difficult to grasp. But in order to reproduce the results of quantum mechanics and to avoid the Bell's proof sort of disqualification, the spin orientation of particle 2 must also depend on the orientation of the Stern-Gerlach device that *particle 1* encounters. Thus, particle 2 is causally affected by our choice of how to orient the device at #1. And since this effect happens instantaneously, the causal signal must travel faster than the speed of light. A non-local hidden variable theory explicitly violates the special theory of relativity.

If we put the results of Bell's proof together with the special theory of relativity, hidden variables in general are ruled out. Bell disallows local hidden variables, and the special theory of relativity disallows non-local hidden variables.

We should put this result in terms of subjective and objective probability, since that will be the key to the issue of appearance and reality. It amounts to a demonstration that the quantum mechanical probabilities associated with spin orientation are objective probabilities. They reflect an inherently indeterminate situation in the nature of things. Quantum mechanics can only give us the probability that particle 1 will be V+, but this is not from a lack of information on our part. It is not a weakness in our way of knowing or an incompleteness in the theory. It is not that some aspect of reality is hidden from us. The probabilities are part of nature. It is all there is to know.

This is clearly relevant to the issue of appearance and reality and understanding the physicist's ability to know how nature is. We will assess this relevance in the next chapter. But first, a brief summary of the basics of quantum mechanics will be useful to make sure the subsequent philosophy is well informed by the facts.

SUMMARY OF QUANTUM MECHANICS

Two important concepts distinguish quantum mechanics from the classical physics that preceded it and that still govern our common sense. Both of these concepts are based on experiments. One is the principle of complementarity. The other is the idea of superposition states in nature.

The principle of complementarity in its general form is a guideline for how we must describe nature. Quantum things, that is, very small things, cannot be described as having both exact position and exact momentum, just as they cannot be described as either particles or waves. They must be described as both, but never at the same time. The nature of a quantum thing like light or an elementary particle is of complementary properties that are mutually incompatible yet jointly required. The duality is forced on the quantum mechanical description of nature by the experiments performed on nature. One situation, the double-slit experiment, consistently shows quantum things to be waves. Another situation, the photo-electric effect, consistently shows quantum things to be particles.

Because the two complementary components of the quantum mechanical nature of things are mutually exclusive, the principle of complementarity leads to the idea of properties that are not simultaneously measurable. This is unlike the classical understanding of our relation to nature in which the object to be measured sits passively while we record any of its properties we choose. In the quantum mechanical model, we still have a choice as to what property to measure, but choosing one may preclude access to another. Choose to measure position, for example, and the exact value of momentum is lost. Choose to measure spin orientation along the V-axis, and the value of spin orientation along R or L is lost. But it is not our fault. It is because the quantum thing does not have an exact momentum when it has an exact position. Nor does it have an exact spin orientation along R or L when it is determinately spin-up (or down) along V.

To say that a quantum thing does not have an exact momentum, or that its spin orientation is indeterminate, is to say that it is a superposition of the possible values of momentum or spin orientation. Like a non-partisan voter, things could genuinely go either way, though one way may be more likely than another. Non-partisans can lean to one side or another. The mechanism of quantum mechanics, the details of the theory, describe the superposition states of quantum things in terms of the likelihood of going one way or another. The state function describes an object in terms of all possible values of a property (position, for example) and the probability for each value, that it will be the result if the property is measured.

We never observe superposition states in the natural things that are big enough for us to observe. Nor can we observe probability or probabilistic behavior in a single event. Just as with dice, where a single roll always shows one value or another but never the probability of any value, each quantum measurement dissolves the superposition and shows us only one of the possible values. So it is not just that there *is* no direct evidence of superposition; there *cannot be* direct evidence of superposition. There is though, compelling indirect evidence of superposition. Bell's proof demonstrates that a quantum thing cannot have a determinate value of the property spin orientation before that property is measured. Only by treating it as a superposition of spin orientation values, having no one value in particular but having each in some probability, can we make predictions that are consistent with the results of experiments.

Quantum mechanics is in an odd situation with respect to appearance and reality. Though its claims are firmly rooted in experiments, it describes the quantum world in terms that do not apply to anything we experience in nature. Nothing we observe demonstrates mutually incompatible and jointly necessary properties. Nothing in our experience, that is, seems to fall under the principle of complementarity. Nothing in our experience is in a state of superposition. In the world we observe, everything is definitively here or there, moving at one speed or another, pointing this way or that. We cannot even imagine what a superposition of properties would look like. Try to visualize a compass needle that is, at one instant, both pointing left and right, yet pointing in neither direction.

Bohr had quantum mechanics in mind when he excused physics from the task of describing reality. Now that we have quantum mechanics in mind we can reasonably ask if Bohr's claim is warranted.

Chapter 7
QUANTUM MECHANICS
AND REALISM

—————————————— • ——————————————

> There is no quantum world....only an abstract quantum description. It is wrong to think that the task of physics is to find out how nature is. Physics concerns what we can say about nature.
>
> Niels Bohr

I have added some valuable context to Bohr's claim about the appropriate task of physics and the issue of appearance and reality. Perhaps it was disingenuous to have left the reference to the quantum world off the earlier presentations of the Bohr quote, but I did it to avoid confusion and misunderstanding. The misunderstanding would be of quantum mechanics and the quantum world. You need to know a bit about quantum mechanics to fairly interpret sweeping claims about the quantum world. Now that you do, you can. The confusion would be between the two separate claims that Bohr is making in this expanded passage. Perhaps the two are related. Perhaps one is the reason to believe the other. This is what we need to sort out in the present chapter. But Bohr is making both a metaphysical claim about the way some aspect of nature *is*, and an epistemological claim about our ability (or inability) to *know* about some aspect of nature. The metaphysical claim is that there is no quantum world. The epistemological claim, that we cannot know about the quantum world, comes close to being inconsistent with the metaphysics. If we really cannot know anything about the quantum world, how can we be sure it does not exist? If the epistemic limitations are true, then perhaps a form of quantum agnosticism is more in order than Bohr's professed atheism. This too is the sort of question to be addressed in this chapter on quantum mechanics and realism.

Both the metaphysical and epistemological points made by Bohr are prone to extravagant misinterpretation. Consider the metaphysics. What does it mean to say that there is no world at the level of quantum mechanics? All of the basic constituents of matter—protons and neutrons and electrons, the components of atoms, and photons, a principle conveyance of energy and information—are squarely in the quantum world, the world that allegedly does not exist. If big things like tables and chairs and you and I are made of little things like atoms, yet atoms are of a world that does not exist, what becomes of us? If there is no quantum world, how is there any world at all?

We can see from our participation in the mechanics of the quantum theory, that Bohr must mean that properties like spin orientation, momentum, and position, are indeterminate in the quantum world. Things in the quantum world exist in superposition states, composites of incompatible attributes. The indeterminate quantum properties are made determinate by us in the act of observation, by the physical interaction with our big, classical bodies and machines with the quantum things. Thus, the claim is not that quantum things do not exist; rather it is that their properties (like spin orientation) are not independent of us. This claim is certainly more precise and clearer than saying that the quantum world does not exist.

The epistemological point also wants clarification, in particular as to its motivation. What is it about quantum mechanics that should convince us that we cannot know how nature, that is, the quantum world, is? If there is nothing there to know, then of course we cannot know much about it. But this is no limitation on the task or ability of physics. If the metaphysical point is supposed to motivate the epistemology, then we must know the metaphysical point to be true. And then there is no more we can say about nature, because there is no more to say.

I do not mean to either endorse or confute Bohr's claims with these preliminary comments on the metaphysics and epistemology in his remarks. I mean only to unpack his provocative and perplexing words so we can analyze them in some detail. Both aspects of his claim are allegedly founded in quantum mechanics itself. That is what we want to see for ourselves.

WHAT TO MAKE OF BELL'S PROOF

Putting aside for a moment the possibility of non-local hidden variables, we must say that the quantum particles 1 and 2 that have been produced from the central, zero-spin object, have no determinate spin orientation before they encounter a Stern-Gerlach measuring device. Each is in a superposition state of both spin-up and spin-down along any axis. As a gesture toward Bohr's metaphysical language, we can say that the spin orientation of any quantum object does not exist until we look to see what it is. This is the clear and definite conclusion of Bell's proof.

It is at this point in interpretations of quantum mechanics that things are prone to get out of hand. We hear the sirens of idealism with the song that nothing (in the quantum world) exists unless we are looking at it. There is a natural thrill in somehow throwing the switch on all of reality, but it is not warranted by the results of quantum mechanics. Neither the general foundations of quantum mechanics, nor the particular conclusions of things like Bell's proof, give reason to endorse the alarmist metaphysics in claims to the effect that reality does not exist unless we are measuring it, or that events in the world are entirely indeterministic, just one damn thing after another. Bell's proof is exactly the place to stop this kind of talk. We must be careful to point out exactly what it does prove and what it does not prove.

First of all, Bell's proof has nothing to do with determinism or indeterminism. It has to do with indeterminateness. This is an unwieldy word, but it is necessary to distinguish between the dynamic concept of determin*ism* and the static concept of determin*ateness*. The former has to do with the nature of change and the progression from the state of affairs at one instant to the state of affairs at another. If the way things are at one moment does not uniquely determine the ways things are at the next, then this is a case of indeterminism. But that is not the issue in Bell's proof. Determinateness has to do with the state of affairs at one instant, without regard as to how it came to be, or what it will lead to. If some properties of things are, at the moment, without exact value, then those properties are indeterminate.

Bell's proof is about indeterminateness of spin orientation, but only of spin orientation. It is about a particular property, spin orientation, and it reaches the specific conclusion that there can be no determinate value of that property prior to the measurement interaction. The derivation depends essentially on the specific characteristics of spin, and it is important to realize that it is not a general proof about all properties of quantum particles. It is not about all of reality, whatever that means; it is just about spin. Furthermore, all proofs that locally determinate-value descriptions of the quantum world (local hidden-variable theories) conflict with quantum mechanics and with experiments are specific to individual properties like spin, position, or linear momentum. These proofs can justify only specific claims to the effect that a particle has no position (or no momentum or no spin orientation) independent of the act of observing that property. They may even license similar claims about big things like the moon, that it has no spin orientation or position when no one is looking, though this extension may require more work. But what is important is that no such proof justifies the general metaphysical claim that a particle has no properties whatsoever. Proof that entities lack some properties is not a reason to think that they lack all properties, and it does not threaten their existence or reality.

If we are using electrons in the EPR or the Bell experimental setup, there certainly are some properties of the thing that, according of quantum mechanics, are determinate at all times. The electric charge is one. The magnitude of spin (distinguished from the orientation of spin) is another. And mass is a third. There are no Bell-type proofs of indeterminateness of these properties. These, according to quantum mechanics, can exist even when no one is looking or measuring. Furthermore, these are among the characteristic properties of an electron. Electrons are identified as objects that have electric charge of 1.6×10^{-19} Coulombs, spin of $\sqrt{3}\, h/4\pi$, and mass of 9×10^{-31} kilograms. There is no reason, other than the metaphysical predisposition to make all of reality indeterminate, to say that these properties do not have exact values when no one is measuring them. Quantum mechanics itself does not require, or even suggest, this wholesale indeterminateness.

A careful assessment of Bell's proof leads to the general admonishment to deal with reality on a case-by-case basis. Some properties are determinate, others are not. This is no reason to say that all of reality is observer-dependent or that things do not exist when no one is looking. There is an analogy to relativity that brings this point home. The property of to-the-left is indeterminate; it has no value (neither yes nor no) until we specify the reference frame. My pencil has no intrinsic property of to-the-left. The value of that particular property depends on something else. But we would not say that, because this one property is indeterminate, the pencil simply does not exist.

The results of Bell's proof, what it does prove, are specific. They are also decisive. Bell's conclusion is a clear and definitive statement about the spin orientation of a quantum object, namely that the property is indeterminate prior to observation. The result is not that we simply do not know the spin orientation, or even that we cannot know it. It is not a case for agnosticism about spin orientation. It is a claim about how quantum objects *are* when we are not, in fact could not be, looking. It is about the way things are (or perhaps more accurately, the way things are not), and not about the way things appear. Spin orientation always *appears* in one determinate value or another. It is always *measured* to be either up or down. But now we know that things are not as they appear.

At the beginning of the previous chapter, during the preliminary remarks on probability, we said, "If we maintain that probability is a property of each individual roll of the die, then we participate in a description of nature, an account of how things are, that goes beyond how things appear." Now we see that Bell's proof shows that each individual particle in the EPR experiment is inherently probabilistic with respect to spin orientation. Probability is a property of each particle, even though each appears to be determinate upon measurement. Bell's proof shows that the quantum mechanical description of nature goes beyond how things appear.

A metaphor about microscopes, suggested by Erwin Schrödinger for another purpose, will help bring together the physics and the philosophy of the situation. Our image of the world as we understand it through quantum mechanics is somewhat fuzzy. We describe such properties as spin orientation only in terms of probabilities, and not in the sharp clarity of determinate values. The original EPR experiment, before Bell, was intended to show that the specimen itself is sharp (in the sense of having determinate values) but that our description of the specimen is out of focus. The idea was that it is our own conceptual imaging device (quantum mechanics) that blurs the image by its incomplete understanding of the specimen. But the results of Bell, the demonstration that there *are* no determinate values, shows that the image is fuzzy because the specimen itself is fuzzy. The specimen is not out of focus.

Bell's proof is thus good news for epistemological realism. It extends our knowledge beyond appearance. Physics, in the form of Bell's proof, takes up the task of describing how nature is, even when we are not looking. The Bell results are in fact doubly good news about the ability of physics to know more than mere appearance. For one, we are able to tell whether or not the conceptual apparatus of physical theory is in focus. This awareness and accountability of our conceptual perspective is an essential step to objectivity. And in addition, the Bell proof shows that the apparatus *is* in focus in the case of spin orientation. It is sufficiently focused to give us knowledge about what we cannot directly observe. By providing information about the specimen, Bell also gives us information about our ability to know about nature.

On the two issues of realism, metaphysical and epistemological, Bell's proof is neutral on the former and reason for realism on the latter. On the claim that there is no quantum world, Bell's proof gives no support, since it is limited to a few specific properties and gives no warrant for generalization to the whole quantum world or all of reality. On the claim that physics cannot describe how nature is when we are not observing it, Bell's proof is that, for some aspects of nature, we can.

It is worth reviewing some of the challenges to realism as suggested in Chapter 2 to show that quantum mechanics and Bell's proof are not only compatible with some aspects of realism, they provide the proof to meet the challenges. At first glance, Bell's proof seems to be evidence for anti-realism. It demonstrates, for example, the indelible physical influence a human observer has on the system being observed. We create the appearance by making determinate spin orientations out of superpositions. In this sense, we force the hand of nature in the act of observation, rather than the other way around.

Furthermore, Bell's analysis of the EPR experiment shows an essential interconnectedness of the system. Even when particles 1 and 2 are separated by a large distance, the properties of one depend on those of the other. There is this inescapable non-locality, this holism. The system exists as a whole rather than as isolated parts. We cannot isolate particle 1 from particle 2, nor can we isolate the observer from the observed. They are all part of a unified system. We are inseparable from nature and no longer can we envision nature as an independent object just waiting about for us to know.

The Bell proof thus articulates the challenges to realism, both epistemological and metaphysical. But a careful analysis shows that the details of the proof in fact support a kind of epistemological realism. It is not just that Bell's proof does not support anti-realism; it provides a positive case for epistemological realism by demonstrating our ability to know more about nature than the way it appears. There is no denying that we create the appearance by influencing the outcome of observations, but that does not mean we influence all aspects of the reality of nature, nor that we cannot know about some aspects. If we keep track of the details of our own influence, we can reconstruct what must have been nature's influence, those aspects of reality. And this is just what the Bell proof does, at least for one aspect of reality, spin orientation. We now know, we have proof, that our observing changes a spin orientation superposition into a determinate state. We can use the features of appearance, namely the probability of different outcomes being 1/4 rather than 1/3 or greater, to infer the reality of the situation. That is exactly the goal of epistemological realism.

This is not a wholesale proof of epistemological realism. It is not a claim that we can know everything about nature. The Bell proof is specifically about spin orientation. It allows us to say we can know about this particular aspect of reality. It is support for a topic-specific realism.

THE QUANTUM/CLASSICAL DISTINCTION

It is certainly unfair to hold Niels Bohr accountable to the results of Bell's proof. Bohr died before Bell published. But what is important here is not the biography of thinkers or the history of ideas. It is the ideas themselves. The anti-realist ideas for which I have chosen Bohr as the spokesperson, the ideas that are usually judged to be vindicated by Bell's proof, are in fact undermined by Bell's proof.

In the interest of fairness it would be good to reflect on the general quantum mechanical theme and see how the ideas of Bohr's comments fit. This sort of reflection can begin with a careful assessment of how quantum systems differ from classical systems.

In a classical system, there are no complementary properties. There is no principled uncertainty about the properties of things, and no values that cannot be simultaneously measured. Any uncertainties in our measurements of the aspects of nature are there because of human limitations. They are the results of our clumsiness or weakness of vision. Our machines are simply too imprecise because of technological problems or the pragmatic limitations of insufficient funding or not enough energy. In principle, that is, by classical laws of nature, all of these limitations could be made arbitrarily small. These classical uncertainties have no natural lower limit.

Classical properties are always determinate and compatible. The world we experience of large, observable objects certainly seems to be a classical world. The mass, position, spin orientation, and so on of the large-scale things around us seem to have exact values and the value of one does not interfere with the value of another. Some interpretations of the situation will argue that no system is genuinely classical, not even the everyday objects we observe and push around. According to such an account it is just that the non-classical effects are too small to notice unless the objects themselves are small. We will deal with this sort of interpretation later in the chapter. For now it is important to clarify the conceptual difference between a classical system (if there are any) and a quantum system.

Quantum systems, we know, can exist in superposition states. Furthermore, some pairs of properties, like horizontal spin orientation and vertical spin orientation, or position and momentum, are inherently incompatible. It is not just that they cannot be measured at the same time. They cannot exist at the same time. The wave/particle duality of quantum systems, the complementarity of position and momentum and the Heisenberg uncertainty principle, are results of the experiments and principles of quantum mechanics, not of the properties of our participation in observing a quantum system. These are descriptions of the system itself, not of our interaction with the system. But, since these properties of the quantum system never show up on classical measuring devices, the quantum system cannot be directly observed. None of the distinctively quantum mechanical features of the world are amenable to direct observation. The world never appears to us in a quantum state. While quantum mechanics describes things in terms of indeterminateness, superposition, and complementary properties, these things never appear.

There is perhaps something suspicious about a theory that builds in a principled distinction between what is observable and what is unobservable, and then packs the exciting and outlandish events into the category of the unobservable. It is as if I told you that I can dunk a basketball like Michael Jordan himself, but for some reason, nerves maybe, whenever anyone watches me, my legs go weak and I can barely leave the ground. You would not believe me, nor should you. And yet, if we agree that observations are classical and that quantum things can be in superposition states, it seems that for some reason the act of observing these things knocks the superposition right out of them. The act of observation changes the quantum system into a classical system. So anytime you look, you see classical properties. How then can observations, which are always classical, be evidence of a quantum reality?

There is a crucial difference in the link between evidence and reality in this interpretation of the quantum mechanical account of things and in the classical, pre-quantum science. Classical physics distinguishes between what is directly observable and what is not. It also recognizes that the act of observation will have an effect on the thing being observed.

Light bounces off things you are looking at, and a thermometer heats up (or cools down) the things being measured for temperature. But in classical physics these effects can in principle be made arbitrarily small. The system after observation can, in principle, be made as similar as we want to the system before measurement. It is this continuity between the way things are observed to be and the way they really are that warrants the use of observation as evidence for claims about the unobserved (even unobservable) reality. With quantum physics though, the influence of the act of observation produces an abrupt, discontinuous change in the system. Things are different *in kind*, not just in degree, between when they are unobserved and when they are observed. The wave function, according to quantum mechanics, collapses from a superposition of possible values to a single, determinate value. The effect of observation cannot be reduced. It changes the system from one kind of thing that we can never observe, to another kind of thing that we can.

How can the observations, which are always classical, be evidence for a quantum mechanical system? This challenge is made even more serious with the realization of the indeterministic aspect of measurement on a quantum system. According to quantum mechanics, the probabilities in the state function of a quantum system are irreducible, objective probabilities. A system that is an even superposition of spin-up and spin-down, that is, with P(spin-up) = 1/2 and P(spin-down) = 1/2, will, when measured, be entirely spin-up or entirely spin-down. Nothing in nature determines which of these two possibilities will be realized. The outcome is, in the nature of things, random. The results of individual measurements are in this sense indeterministic. The state function does not determine the outcome of the measurement. How in the world then can we claim that measurements are evidence of the quantum state function?

THE COPENHAGEN INTERPRETATION

At this point, Bohr's epistemological claim seems to make sense. We cannot use classical observations as evidence for claims about the quantum reality. Classical observations are no more justification for the reality of quantum conditions like superposition and indeterminateness than would your observations of me struggling to jump high enough to brush the net be reason to believe that I can dunk a basketball. So, Bohr concludes, to avoid the slide into the irresponsible dogmas of mysticism, physics should not even try to prove things about the way things are at the quantum level. The state function is a very useful tool for organizing our thoughts and making predictions about the observable world. As a model, a way of thinking, it is of tremendous value. But that is all it is, a model, a way of thinking. At least that is all we should make of it. We use it, and do our calculations under the supposition that quantum things behave as the state function describes. But that is no reason to say that the quantum nature really *is* as the state function describes.

This attitude toward scientific theory, that descriptions of unobservable things should be regarded as useful ways of thinking but not necessarily true, is sometimes called instrumentalism. Theories are instruments, tools for directing our thoughts. And like tools, their value is in being useful, not in being true. We do not even ask if they are true or not, and the task of physics is not to find out how nature is. It is only to facilitate a useful description of nature.

Bohr is an instrumentalist about quantum mechanics, but his attitude is not a philosophical predilection that he brought with him and chooses to impose on quantum mechanics. The claim is that the science itself, quantum mechanics, forces the philosophical view, the instrumentalism.

At this point it is important to distinguish between the facts of quantum mechanics and the interpretations of quantum mechanics. Quantum mechanics itself describes some natural systems in terms of complementary properties, superposition, interference, and probabilities. These are the facts of the science. The physics forces us to make sense of these details, but there may be alternative ways to do it, alternative interpretations. What we are up to here is to see which interpretation is most plausible in light of the physics of quantum mechanics.

Bohr's instrumentalism is a central feature of what is called the **Copenhagen Interpretation** of quantum mechanics. The interpretation was developed from the ideas of Bohr and Heisenberg in Bohr's institute for physics financed by the Carlsberg Brewery. (No kidding. Read all about it in Gamow's *Thirty Years that Shook Physics*.) Even though it is quoted as the received view, the party line among practicing physicists on the appropriate attitude towards the quantum theory, the Copenhagen interpretation itself comes in a variety of versions. It is based on four assertions:

1. *Quantum mechanics is complete.* The probabilities, superpositions, and indeterminate properties in the theory cannot be supplemented by more information (hidden variables) that would reduce the probabilities to single, determinate (classical) values. Classical properties cannot be used to describe the quantum state.

2. *All observables are classical.* Anything that we can observe and measure will show itself with classical properties.

3. *The task of science is to describe relations between observables.* What separates science from mysticism and dogma is the insistence on observable evidence. We can speculate about an unobservable reality, but that is not science.

4. *We cannot ascribe any reality to the quantum state-function or the quantum world.*

The fourth statement apparently follows from the other three. The first is motivated by results of Bell's proof, and so Bohr's conviction was somewhat ahead of its time. The second claim is simply a report of what we experience in the world.

The Copenhagen interpretation warrants the same scrutiny as the Bell proof. Again we must be careful of the alarmist slogans, that there is no quantum world, or that there is no reality when we are not looking, or that measurement (that is, people) create or determine the reality of nature. These, and the more sober Copenhagen interpretation itself, must be compared to the facts about the quantum theory.

First of all, there is no evidence that *all* properties of a quantum system are indeterminate before they are measured. Things like electric charge, mass, and spin have no Bell-type proofs that give us reason to believe that their classical determinateness is inappropriate to a quantum description. Thus, the part of the first premise of the Copenhagen interpretation that claims that *no* determinate classical values can be applied to a quantum system is an overstatement, misleading at best, downright false at worst.

Second, the Copenhagen interpretation as stated in the four basic claims, is inconsistent in the way suggested in the opening paragraphs of this chapter. The first claim is a definitive description of the way the quantum world is. The fourth is a declaration that we cannot know anything about the way the quantum world is. If we cannot know how nature is, then we cannot know how it is not, either. Statement one, and the results of Bell's proof, regard indeterminateness and superposition as facts of nature, the way things are at the level of a quantum system. This is contrary to the instrumentalist attitude of making no claims about unobservable facts. It is inconsistent with instrumentalism, and so it surely cannot function as proof of instrumentalism.

Again, the problem has developed from an unwarranted leap from particulars to a sweeping generalization. What quantum mechanics *does* tell us about our ability to know about nature is best described in terms of specifics. Some properties, like spin orientation, position, and momentum, we now know are indeterminate in quantum systems. Following a kind of instrumentalist attitude about categories and properties, we can say that these properties, while useful in describing large, classical objects, are cumbersome and inefficient in describing very small things. Live and learn. We of course choose what properties to attend to in describing the world, just as we choose what language to use, what geometry to apply, and what measurements to perform. The pragmatic lesson to be learned from quantum indeterminateness is that some of the properties that are useful descriptive handles on classical events are not so useful on quantum events.

All this means that the quantum world is quite different than the classical world of big things. But it does not mean that the quantum world does not exist. The progress in physics that this represents is an unusual kind of progress. Science usually advances by unifying seemingly disparate phenomena under a single law or within a single explanatory model. One of Newton's achievements was to unify the terrestrial and celestial worlds under a single mechanical description. The laws of motion on the earth are the same laws of motion for planets and stars and the moon. Realizing this was a real stroke of progress. Realizing that magnetism and electricity are just different manifestations of the same underlying phenomenon was also progress. And a unified field theory, describing all kinds of motion in terms of a single force, will also be progress, if it ever happens. But quantum mechanics has the opposite effect. Where before we assumed that the behavior and properties of the very small constituents of matter would be just reduced versions of the properties of big things, we now realize that there is a difference of kind. The world of the very small is dramatically unlike the world of the large. The quantum world and the classical world cannot be comfortably unified under a single set of descriptive properties. Coming to terms with this disunity in nature is, in the case of quantum mechanics, progress.

Some interpretations of quantum mechanics deny this disunity, but if we insist, as does the Copenhagen interpretation, on using classical properties like position and momentum to describe quantum things, then we must be willing to deal with the descriptive inconvenience of superpositions. At the quantum level, these properties are indeterminate until measured, that is, until they make a mark on the classical world. Classical properties like being-a-wave and being-a-particle are more complicated at the quantum level than the classical, in that they are two-place relations. A quantum object is a wave with respect to some kinds of interactions (double-slit, for example), and is a particle with respect to other kinds of interactions (photo-electric effect). Being a particle is an intrinsic, one-place property in a classical description. A classical basketball is a particle, period. The property of being-a-particle (or being-a-wave) is no more definitive in quantum mechanics than being-ten-meters-long (or being-nine-meters-long) is definitive in the special theory of relativity.

It is one thing to accept that there is a distinction between the way things work in the quantum world and the way they work in the classical world, but it is another thing to genuinely understand the distinction and the relation. Usually, the accomplishment of understanding is unification. We understand the orbit of the moon now that we know it is just another case of gravity. I understand Bob's illness when I realize that his is a case of chicken pox, and that is just a kind of virus. And so on. But on the border between the quantum and classical worlds, we understand things insofar as we can clearly separate them.

A quantum object like particle 1 in the EPR or the Bell experiment is in a superposition state until its spin orientation is measured. After the measurement it is definitively spin-up of definitively spin-down. The quantum state function of the object evolves smoothly and deterministically from the instant of its creation to the instant of measurement. Then it abruptly changes. The measurement brings about a sudden, discontinuous, and indeterministic collapse of the superposition to a single value. Only the act of measurement, making a record of the value, changes the quantum superposition into a classical value. Other physical interactions between one quantum object and another leave superpositions as superpositions. What is so special about measurement?

THE MEASUREMENT PROBLEM

The problem with measurement in quantum mechanics, and in particular the Copenhagen interpretation, is that we do not understand it.

Observations and measurements are a special kind of interaction in any context, quantum or classical. They are physical interactions that leave a record of information that we can understand. For classical things, whether or not an interaction amounts to an observation is pretty much a contingent, circumstantial thing. An observation is a regular sort of interaction that takes place between the object and a person, or at least with something in a chain of interactions that eventually gets to a person. But in quantum mechanics and its most prevalent interpretations, observations and measurements are by nature a different kind of interaction than others. They have a radically different effect. Observations and measurements collapse the wave function and resolve the superposition to a single, determinate value. Other interactions do not. An electron in an atom is constantly interacting with the nucleus. This is how the electron stays in orbit and the atom holds together. But the electron continues in its superposition state, with its position indeterminate. Furthermore, the state function of the atomic electron evolves in a smooth and deterministic way. There are no sudden abrupt changes. Conditions at one moment exactly determine conditions at the next. But if the electron interacts with a measuring device, something that informs us of where it is, the state function collapses. Its change is abrupt and indeterministic. There is no exact telling which of the possible possibilities will be the measured result.

What is so special about measurement? What is it about a measurement interaction that causes the collapse of a state function? And how does a measurement cause the collapse? What kind of interaction will change a quantum system into a classical system? That is, what is the factor that makes an interaction count as a measurement? Is it a matter of size, such that interaction with anything big, like a Stern-Gerlach device, will bring about collapse? Does it require a human component in the sense that only a conscious awareness of the state function brings about its collapse? And what happens to the other values of the property measured, the ones that do not show up?

This nest of questions is at the heart of the measurement problem.

The issue is not the spooky action at a distance between two objects, as in the EPR experiment. The issue is the spooky behavior of just one object, particle 1, say, in the snap change from superposition to spin-up (or down). Local hidden variables would have solved the locality problem. If particle 2 was spin-down all along, then there was no action at all from particle 1, neither spooky nor at a distance. Local hidden variables would solve the measurement problem too. If particle 1 is spin-up all along but we just do not know about it, then the act of measurement is a change in our knowledge but not a change in the thing itself. With a determinate spin orientation there is no superposition to collapse and no abrupt physical change that takes place during measurement. It is just like walking into the restaurant and seeing that Bob is there. The probability clicks from 1/2 to 1, but it is just a subjective probability. The change is in my head, not in Bob's body. If there are hidden variables in the quantum system that give determinate values to properties before they are measured, then there is no measurement problem. The abrupt change from probabilities to definitive values is in our heads, not in the quantum thing.

But there are no local hidden variables. This is the result of Bell's proof. Thus we are forced to deal with the measurement problem. It is certainly a problem for the Copenhagen interpretation, which regards the indeterminateness and superposition of quantum states as basic, irreducible features of nature. The Copenhagen interpretation draws a sharp distinction between the quantum world, where these sorts of things happen, and the classical world, where they do not. What marks this distinction? Is it just a matter of size? Are big things classical and little things quantum mechanical?

There are experimental reasons to say that size is not what distinguishes a quantum system from a classical system. Size alone is not what makes a measurement a measurement and causes the collapse of the wave function. One such experiment involves Stern-Gerlach devices to measure spin orientation along the vertical axis (V) and along the horizontal axis (H). The results of the experiment are that the small quantum thing can interact with the large classical device and emerge still in a superposition state, that is, still a quantum thing. Only if a record is made of its passage through the device will the state function collapse and the spin orientation become determinate. Thus, it is not interaction with something big that collapses the state function and marks the change from quantum state to classical. It has something to do with making a record of the event.

Here are the details of the experiment. They clarify what we mean by "making a record." The first step is to prepare a supply of particles that are in a superposition between spin + and spin − along the horizontal axis. Do this by sending a random stream of particles into a Stern-Gerlach device that is aligned along the *vertical* axis. The particles come out in two groups, the spin-up (V+) and the spin-down (V−). Block the V− exit so that only V+ particle emerge. Now the beam of particles is pure V+, which means that each particle is a 50-50 superposition of H+ and H−. And sure enough, send the beam into a horizontally oriented Stern-Gerlach device and half of the particles come out the H+ exit, and half come out H−.

This first step in the experiment is shown on the top line of Figure 7.1.

The question relevant to the measurement problem is this: In the interaction with the horizontal Stern-Gerlach device (the second device encountered by the beam), did the H-superposition collapse? If it did, then each of the emerging beams, the H+ beam and the H-beam, will be de-

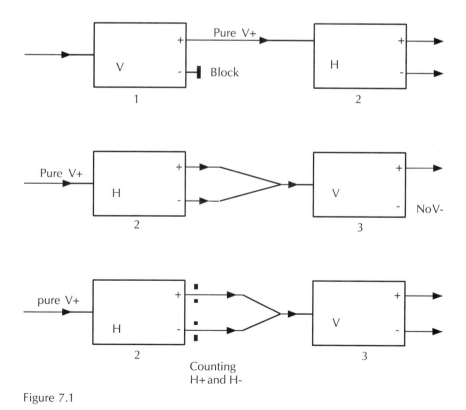

Figure 7.1

terminate values of spin orientation in the horizontal direction and hence *indeterminate* in the vertical direction. Spin orientation along different axes are incompatible properties. Like position and momentum, if one is determined the other cannot be. That is why the original, determinate V+ beam is a superposition of horizontal spin orientation. Similarly, each determinate H+ and H− beam would be a superposition of vertical spin orientation. Each would be a pure superposition of V+ and V−.

If the horizontal Stern-Gerlach device has collapsed the state function, then sending both the H+ and H− beams into a third Stern-Gerlach device, this one vertically aligned, will result in both V+ and V− particles emerging. That is, the vertical Stern-Gerlach device will resolve the vertical superposition in the determinate horizontal states, and measure both V+ and V−. But this is not what happens. As long as no record has been made of the particles emerging from the horizontal device, the vertical Stern-Gerlach measurements finds only spin-up. This is shown on the middle line of Figure 7.1. It is as if the beam entering the third Stern-Gerlach device is pure spin-up, and hence in a superposition of horizontal H+ and H−. It is as if the pure V+ beam passed through the second Stern-Gerlach device but was in no way altered by the encounter. Pure V+ in; pure V+ out. As long as no one looks at the results of the horizontal device, it does not collapse the state function. The most accurate description of this is to say that the composite system, particles plus Stern-Gerlach device, is in a superposition state of H+ and H−. The particles do not exit either the H+ opening or the H− opening. Each comes out of both, yet neither. The Stern-Gerlach device indicates neither H+ nor H−. It indicates both, yet neither.

But if any record is made of the particles emerging from the horizontal device, if a count of H+ and H− is kept, then the combined beam, when sent through the vertical Stern-Gerlach device, produces equal numbers of V+ and V−. This, perhaps, is the spooky part. It is shown on the bottom line of Figure 7.1. If we look to see how many, or just what percent, of the particles come out of the second Stern-Gerlach device H+ and how many H−, then the state function collapses. It is not interaction with a large, classical object like the Stern-Gerlach device that causes the collapse. It is the record of the interaction that does it.

Some people find this to be not just spooky but down right absurd. The measurement problem is often cited as a kind of *reductio ad absurdum* for the Copenhagen interpretation of quantum mechanics. The interpretation relies on a distinction between quantum and classical systems, but it does not know where or how to draw the line. This failure allows some crazy consequences, crazy enough, perhaps, to warrant an alternative interpretation of quantum mechanics.

The craziest, or at least the most famous, consequence is Schrödinger's cat. This is a thought experiment intended to reduce the Copenhagen interpretation to an untenably absurd conclusion. Its point is that no line can be drawn between a small quantum object in a superposition state of spin-up and spin-down, and a big cat in a superposition state of being dead and alive. But we know that cats cannot be somehow both dead and alive. So either the quantum object cannot be in a superposition state either, or we need to know where and how to draw the line between things that can be in superposition and those that cannot.

A version of the experiment is easily set up with the tools at hand. Inside of a sealed, soundproof box, the Stern-Gerlach device for particle 1 of the EPR experiment is attached to the trigger of a poisonous gas dispenser. The attachment is such that if particle 1 measures spin-up, the gas is released. If particle 1 measures spin-down the gas is not released. There is a cat in the box too. It was alive and well when we put it in and sealed the box. Particle 1 enters the Stern-Gerlach device as a superposition of spin-up and spin-down. As long as no one looks in to record the measurement, particle 1 will always be in a superposition. That is, the Stern-Gerlach device measures neither up nor down. It too is in a superposition of the two. So the gas dispenser is in a supposition of releasing the poison and not releasing the poison. So the cat is neither dead nor alive. It is in a superposition of being dead and being alive.

There is no way, given what we know about biology and about cats, to make sense of a superposition between dead and alive. As soon as we open the box of course, a record of events is made and the superposition dissolves. We will see a dead cat or a living cat. Superpositions occur when no one is looking because observation collapses the state function. Thus no observation can directly falsify the claim that the cat is neither dead nor alive (yet somehow both) until it is observed.

This is a challenge to the Copenhagen interpretation. We have no clear idea what happens during measurement to collapse the state function, and so we have no idea what counts as a measurement. And yet measurement plays such an important role in the Copenhagen interpretation. It is not just an epistemologically distinctive event in that it adds to our knowledge of nature. It is a metaphysically distinctive event, in that it changes things in nature from being indeterminate to being determinate. The Copenhagen interpretation takes this curiosity in stride, as it accepts non-locality and superpositions themselves as basic features of nature. The measurement problem, and its graphic display in Schrödinger's cat, make quantum mechanics and the Copenhagen interpretation even more amazing. They point out something, that we really ought to understand but that we do not yet understand: measurement. There is no reason to conclude that we cannot understand it, ever.

ALTERNATIVE INTERPRETATIONS OF QUANTUM MECHANICS

Dissatisfaction with the Copenhagen interpretation and its cavalier dismissal of the measurement problem has led to a variety of alternative interpretations of quantum mechanics. It is important to realize that none of these alternatives (except Bohm's theory) question the basic quantum theory. None (again, except Bohm's theory) cast doubt on the points raised in the last chapter. They differ from the Copenhagen interpretation and from each other in terms of what to make of quantum mechanics. Looking at these alternatives then is not so much analyzing physics as it is a look at what has come after the physics, the metaphysics.

We will look at four alternative interpretations, and those only briefly. They are chosen because of their high profile in the literature and because they help to illuminate what physics itself has to say about appearance and reality.

The Many-worlds Interpretation

The first is called the many-worlds interpretation. It addresses the measurement problem by focusing on the question of those values in a superposition that do *not* show up in the measurement. It is a way of understanding the collapse of the state function. If a quantum object in supposition of spin-up and spin-down is measured to be spin-up, what has become of its spin-down component? The way we have been talking so far, it simply disappears. Under the many-worlds interpretation though, the spin-down component does not disappear; it shows up in another universe. The state function does not collapse, it splits into two separate worlds.

At the moment of the two-level measurement, a second universe is created. Or it might be better to say that from one old universe, two new ones are created. In one universe the measurement is spin-up. In the other the measurement is spin-down. We are in one universe or the other. In this case it is the one that got the spin-up component. Or perhaps we should say that there is a version of each of us in each universe but the different versions cannot interact in any way. The version of you in the spin-up-measuring universe can have no awareness of other versions of you in other universes.

Every actual measurement brings about a whole universe for every possible value of the property being measured. It may seem ironic that this account is proposed to avoid the spookiness of the collapse of the state function, but now measurement is seen as a branching of universes rather than a collapse of the state function. Measurement is a selection process rather than a causal process of collapse.

The details of the many-worlds interpretation can be worked out to fit the experimental results and the basic principles of quantum mechanics. Nonetheless, there are reasons to doubt that it amounts to any progress over the Copenhagen interpretation. One reason is obvious. Whatever explanatory problems the many-worlds interpretation might avoid, it does so at a staggering metaphysical cost. Consider how many measurements have been made since the beginning of time, and how many possible values each has had. That is how many universes there are presently. Of course, we, or rather, any one version of us, can observe only one universe. All the others are said to exist in a way that is strictly unobservable. But perhaps that does not bother you.

A second problem with the many-worlds interpretation is that it does not really address one of the main aspects of the measurement problem, namely what counts as a measurement. Sometimes, in estimating the number of universes entailed by the many-worlds interpretation, people figure the number of particles in the universe and the number of interactions they would have had. Each interaction marks a branching off of new universes. But should we figure the number of *interactions* or the number of *measurements*? And what is the difference between the two? This is a central part of the original measurement problem, and insofar as the many-worlds interpretation fails to address it, the interpretation fails to help.

The Consciousness Interpretation

Another alternative interpretation of quantum mechanics focuses on exactly what the many-worlds interpretation neglects, the distinguishing factor between measurements and other interactions. It proposes that the key ingredient in a measurement is human consciousness. By this interpretation, Schrödinger's cat is in fact in a supposition state, between being dead and being alive, until we look in and become aware of the cat. The collapse of the superposition coincides with the achievement of knowledge, and only things capable of knowledge, that is, only humans, are capable of causing the collapse.

The details of the interaction between physical object and the mind of the observer are vague. But we can be clear that this is not a reheated version of subjective probabilities, where the act of measurement only completes our previously partial knowledge of the situation. On that description, measurement has no profound effect on the object itself since its properties are already determinate. The subjective probability description relies on local hidden variables and is once and for all untenable in light of Bell's proof. But this new interpretation that invokes consciousness as the key collapsing factor of measurement, involves a two-way interaction between the physical object and the mind. The object affects the mind in some way; that gives us knowledge. And the mind affects the object by causing a collapse of the state function. Some properties are indeterminate until affected, in some unspecified way, by a human mind. Only when we look do these physical properties have determinate values.

Note the difference between this relation between properties and persons and the relation between properties and reference frames in the special theory of relativity. The special theory of relativity has nothing to do with people. It is about reference frames, whether or not there is a person on hand to observe and gain knowledge. On this consciousness interpretation of quantum mechanics though, some properties take on values only when there is a person observing.

There is a kind of metaphysical dualism presupposed by this consciousness interpretation of quantum mechanics. There are two distinct kinds of stuff in nature, mind and matter. The interaction between the two takes place in an act of observation or measurement. Some properties of matter are indeterminate until the matter interacts with a mind. The interpretation is short on details about this interaction, saying only that it happens and the result is a collapse of the superposition state function. There are superpositions, as between spin-up and spin-down, in the physical world of matter, but none in the mental world of measurement. We are always aware of determinate values.

It is ironic that, in this interpretation, there is physical superposition but not mental superposition. It is ironic because, by common sense, it is mental states that seem prone to ambiguity and seem closer to the notion of superposition. We are used to mental ambivalence. Do I want to go to the movies? Well, I do and I do not. Do you understand the new tax laws? Well, yes and no. But in the consciousness interpretation of quantum mechanics it is the mind that gives a definitive nature to the physical world.

This interpretation raises another opportunity for things to get out of hand in drawing consequences from quantum mechanics. There might be an urge to sum it all up with a slogan or two. The world is thoroughly subjective, we might say, or consciousness is an essential element in everything in nature. But as with most exciting slogans, these are misleading. It is important to remember that this consciousness interpretation is not quantum mechanics proper. This is a philosophical addition to quantum mechanics, an interpretation, and only one of a large selection of interpretations. Thus, it would be wrong to say that physics in general or quantum mechanics in particular indicates a necessarily subjective or mind-dependent state of nature.

It is also worth noting that the consciousness interpretation itself is quite speculative and vague. The important interaction between mind and matter is unexplained. There is no clue as to the mechanism by which consciousness affects physical objects and causes the collapse of the state function. The consciousness interpretation does not offer much progress in understanding quantum mechanical measurement, since it explains one mysterious phenomenon (the collapse of the state function during measurement) in terms of an equally mysterious phenomenon (the interaction between mind and matter).

It's All Quantum Mechanics

The previous interpretations try to come to terms with the distinction between the quantum world and the classical. This next one denies the distinction altogether, and thereby aims to dispel the measurement problem. The idea is to unify the description of nature under a single theory that applies as well to quantum-sized things like electrons as to large, observable things like cats and planets and Stern-Gerlach devices. Since we know from Bell that we cannot use a classical, superposition-free description of a quantum-sized system, the only option for unification is to use the quantum theory on the macroscopic world of things we observe. Everything, in other words, is a quantum mechanical system.

If this works, and there will be some difficulties, it will avoid the measurement problem. If there is no change in kind, from quantum system to classical, from superposition to mixture, then there is nothing distinctive about measurement. It is just a kind of interaction. We do not have to explain the collapse of the wave function, because there is no collapse. If the system is a superposition before measurement, it is a superposition after measurement. The system, in this case, includes all the interacting objects, large and small. So the particle, the electron or whatever, and the Stern-Gerlach device are in superposition of spin-orientation at all times. The macroscopic object, the Stern-Gerlach device, or the cat whose life depends on it, or the person who notes the results, must be in a superposition state.

There is considerable aesthetic appeal in this approach with its unity and tidy continuity between large and small. The problem is that it defies our everyday observations. We observe determinate states, not superpositions. Quantum mechanics makes some wild claims about the systems it describes, none of which show up in our observations of cats or Stern-Gerlach devices. But in fairness, recall that the special theory of relativity made some wild claims, that length and time duration are relative to speed, for example. These apply to all systems, fast and slow, yet we do not see them in our day-to-day observations. The relativistic effects are there in all systems, but they are *unnoticeably small*, for all but very high speeds. The special theory of relativity applies to everything, but its unique effects, its differences from Newtonian mechanics, are only significant at very high speed.

Perhaps it is a similar situation in quantum mechanics, that it applies to everything but its unique effects like superposition are significant only for very small objects. Quantum mechanics applies to large objects and its effects are present, they are just, for some reason, insignificant and undetectably small. On this interpretation, describing the particles leaving a Stern-Gerlach device as a mixture, that is, each collapsed into a determinate state, is a false but adequate account, just as Newtonian mechanics is a false but adequate account of objects at slow speeds.

We need an explanation of why the distinctively quantum mechanical effects are insignificant in larger objects. The superposition that is essential to the description of a single electron becomes insignificant and hence inessential to the description after measurement. Measurement involves interaction with a large object, or more importantly a *complex* object composed of many parts, many atoms. Apparently, as information is amplified from the tiny, elemental electron to the vast Stern-Gerlach device, the effect of superposition is diluted through the many parts of the measuring device. An analogy to the effects of thermal motion is often used to make this explanation more plausible. In any large, composite object, the parts, the molecules, are in random motion, oscillating back and forth. This is what causes heat. But this motion is undetected in the large object since the energy is diluted randomly in the smaller parts. It is not that the large objects are governed by different laws; it is just that the effects of thermal motions do not show up in the complex object. Perhaps it is similar with the quantum mechanical effects.

There is still a problem. Consider a single particle, an electron, going through the series of Stern-Gerlach devices shown in Figure 7.1. Before encountering the second Stern-Gerlach device, the particle is in a superposition state for horizontal spin orientation. If a record is made of its exit from the second Stern-Gerlach device, this superposition collapses (on other interpretations) or is diluted (on this interpretation). But if no record is made, there is no collapse, no dilution at all. The superposition is intact. In both cases there has been interaction with a large, complex object, the Stern-Gerlach device, so complexity does not seem to account for the different outcomes. There is still something unique about measurement.

Bohm's Theory

David Bohm has proposed another account of the measurement process in which there is no abrupt collapse of the wave function. But Bohm's version is not really an interpretation of quantum mechanics, because it denies such essentially quantum mechanical phenomena as indeterminate states, complementarity, and metaphysical probabilities. It does not interpret quantum mechanics, in other words, it avoids it. That is why I call it Bohm's theory rather than Bohm's interpretation. The idea is to describe nature with a theory that has all the successful experimental results of quantum mechanics and passes the test of Bell's proof, yet does it without the oddities of indeterminateness or indeterminism. As the details of the theory will reveal, there is a catch. The price of determinism and determinateness is a violation of the special theory of relativity. Bohm's theory requires casual signals that travel faster than the speed of light.

Here are the basics of Bohm's theory. Any natural system such as an electron has two real, determinate, coexisting components, a particle and a wave. The particle is a normal, classical particle with a definite, determinate position and momentum at all times. Its trajectory through space is deterministic. The associated wave is a real, physical wave that exerts an effective force on the particle as a water wave can push around a cork. It is not like the abstract quantum mechanical state function. The wave in Bohm's theory never collapses. There is never an abrupt resolution of possible outcomes of measurement, since there is no superposition of values. It is always determined how a measurement will turn out, given the specifics of the measurement setup and process. There is a relation between the Bohm wave and the quantum mechanical state function. If you do not know the position of the particle, but you do know the mathematical description of the wave Ψ (we use the same symbol for the wave in Bohm's theory as in the quantum theory), then the probability of the particle being at a particular point is the square Ψ^2. Note that all talk of probabilities in Bohm's theory is about epistemic probability, a measure of our lack of information, not a metaphysical haziness. Nonetheless, the mathematical similarity in linking probability of position to square of the wave function is what allows the Bohm theory to reproduce the empirical predictions of quantum mechanics.

The property of position of the particle is fundamental, and all other properties are dependent on position. All of the measurements we have been talking about in the discussions of quantum mechanics come down to a measurement of position. Measuring interference, for example, to reveal the wave-like properties of something, depends on noting where on the screen the light or the electron or whatever hits. It is a measurement of position. Measuring spin orientation depends on noting which exit of a Stern-Gerlach device emits the object. Again, the final information is of position.

Bohm's theory describes how the position of a particle determines the outcome of a spin orientation measurement. The position is always determinate and exact, so the outcome of the measurement is always deterministic. Exiting the Stern-Gerlach device through the spin-up (+) exit or the spin-down (−) exit depends on the local Ψ field inside the device pushing the particle one way or the other. The field is different at different points in the device (in more mathematical terms, Ψ is a function of position). If the particle moves into an area where Ψ pushes it toward the spin-up exit, it comes out spin-up. If it come into an area where Ψ pushes down, it comes out spin-down. The outcome depends on exactly where the particle enters the opening of the Stern-Gerlach device. If it comes in a hair high, it will encounter the spin-up influences. A hair low and it goes spin-down.

Suppose a particle is just about to enter a Stern-Gerlach device through the upper half of the entry port. If we let this measurement take place without our interfering, it is a sure thing that the result will be spin-up. But just before the particle enters we rotate the device by 180° about an axis that passes through the center of the entry port and is parallel to the incident trajectory of the particle. This has the effect of rotating the Ψ field as well, since it is created by the mechanism inside the device. The particle *was* heading for an area of spin-up influence, but it is now heading into an area of spin-down influence. The outcome of the measurement will be spin-down. This shows that while spin orientation is a completely determinate property, it is a jointly held property of both the particle and the measuring device. The position of the particle is intrinsic, a property of the particle on its own. Position in Bohm's theory is a local hidden variable. (More in a minute on why it is hidden. What is important now is its locality.) Spin orientation is a non-local hidden variable. It is a *contextual* property in that its value depends on the particle and the context in which it interacts.

There are common examples of contextual properties that we encounter every day. If you toss a ball toward a backboard and ask about its trajectory after it bounces back, the answer will depend on the orientation of the backboard. If the ball goes straight in, it will come straight back. But if someone can turn the backboard, you can throw the ball again in the same direction but have it come back at an angle.

The case of contextual properties in Bohm's theory is a bit more bizarre than bouncing a tennis ball. It gets downright spooky, as you might expect if the theory is to match the empirical success of quantum mechanics. In the EPR experiment, if particle 1 is spin-up, then particle 2 must be spin-down. But suppose we rotate the Stern-Gerlach device just before particle 1 enters it as we did above. If particle 1 *was* in the position to come out spin-up, it will now be set to come out as spin-down. By rotating the device, by changing the context of its interaction, we have changed its spin orientation. We have also changed the spin orientation of particle 2, since it *was* about to be spin-down but is now spin-up. This is an actual physical change of an actual physical thing. It is a change that has happened at particle 2, caused by our actions some distance away at particle 1, and communicated at an instance. This is a blatant violation of that claim in the special theory of relativity that allows no causal signal to travel faster that the speed of light.

The violations of the relativistic speed limit are always hidden, according to Bohm's theory, and this is what makes the theory empirically, if not conceptually, compatible with the special theory of relativity. We could detect and use the faster-than-light signal only if we knew the precise positions of particles 1 and 2. We have to know exactly where they are entering the Stern-Gerlach devices in order to note the difference between what their spin orientations would have been from what they are. But we cannot, in principle, know the exact position of an object. This is an epistemic imprecision, not a metaphysical fuzziness. It is a genuine uncertainty principle. The particle has an exact position at all times, we just cannot know what it is. The uncertainty is a result of our unavoidable physical influence on the object in the act of measurement. Knowing its position always causes some alteration in its trajectory. We change its motion, and change it in a way that information on the particle's original position is lost, or at least blurred.

This uncertainty that plagues the act of measurement is what leads to all the probabilistic results in the experiments of quantum mechanics. In Bohm's theory, it is an explicitly subjective probability.

Bohm's theory is often described as a realist account of measurement, and indeed, it is a safe haven for metaphysical realism. All the properties of natural things are determinate. Measurement is just an interaction between a physical particle and a physical forcefield. Outcomes are all deterministic. God does not play dice. The cost of this metaphysical realism though is a built-in and enforced epistemological anti-realism. The determinate properties, including the fundamental property of position, must be hidden from us, incorrigibly imprecise.

When applied to the situation of the EPR experiment and Bell's proof, this is exactly the opposite interpretation from the one I have suggested. Bell, I claim, demonstrates that the probabilities are in nature, not in our heads, and that indeterminate properties are a fact of life. And it is this demonstration that bears the burden of proof for epistemological realism, at least for the particular property of spin orientation. There is no uncertainty over spin orientation, only indeterminateness.

SUMMARY

There is no denying that the results of quantum mechanics are bizarre and that they present a challenge to our understanding and credulity. The description of the quantum world includes superposition states, complementarity, properties that are inherently indeterminate, and irreducible probabilities. Some events in the quantum world happen without determinism. None of this stuff seems real or natural to us. Certainly none of it is what common sense expects to find in nature. But just because the quantum mechanical description defies our expectations about nature does not mean that it is just a subjective description and is not about the way nature is. If we look for swans, assuming that all swans are white, and we come to a land with black swans, it does not mean that these swans do not exist or that the report of black swans is not a report of reality. It means we are someplace weird where swans are black. The quantum world is someplace weird, but that does not mean it does not exist.

And not even all the quantum swans are black. That is, not all properties are indeterminate and not all events evolve indeterministically. It is true that there can be no local, hidden variables that make the values of spin orientation, position, momentum, and some other properties determinate. These are indeterminate in the nature of things, but that does not mean that all properties are indeterminate. The moral of the quantum mechanical story is that the issue of realism can only be addressed on a case-by-case basis. The nature of spin orientation is different than the nature of electric charge. One is indeterminate before measurement, the other is not. And so our relation to the property of spin orientation is different than our relation to electric charge. With one a determinate value is collapsed out of the superposition when we measure. This is not the case with the other. These differences show that we cannot legitimately generalize about the whole quantum world or all of reality. We must be specific.

This counsel against generalization also applies to claims about determinism. Quantum physics does not force us to give up determinism about all events. According to quantum mechanics, some sequences of events are entirely deterministic, others are not. In fact, within the quantum world, all processes are deterministic. It is only at the transition between the quantum and classical worlds, only when the state function collapses to a single value, that things are not deterministic. The state function evolves deterministically until this event.

The best news about realism to come out of quantum mechanics is Bell's proof. Now we know that spin orientation is indeterminate before measurement. It is not uncertain. It is not that we simply do not know the spin orientation. It is in fact indeterminate when we are not looking. This is a definitive, justified claim about some aspect of nature as it behaves when unobserved. It is a claim of knowledge that clearly transcends appearance, since the objects in the experiment always appear to have, that is, are measured to have, determinate spin orientation. Proving that things are not as they appear is what epistemological realism is all about.

Chapter 8
REALISTIC REALISM

─────────────────────── • ───────────────────────

There is no quantum world. ... only an abstract quantum description. It is wrong to think that the task of physics is to find out how nature is. Physics concerns what we can say about nature.

Niels Bohr

Physics is an attempt conceptually to grasp reality as it is thought independently of its being conceived. In this sense one speaks of physical reality.

Albert Einstein

We have been using Bohr's remarks as a summary statement of anti-realism. There are two parts to his remarks, one about nature itself and the other about an aspect of our relation with nature, knowledge. On the first part, there is no quantum world. On the second, we cannot know how nature really is; we can only know nature as it appears to us.

Einstein's comment is a claim of realism. It too has two parts, one about nature itself and the other about our knowledge of nature. On the first, there is an independent reality, an existing state of nature in all realms of sizes, quantum and classical. Furthermore, we can know about it. We can transcend the appearances and come to know how nature really is.

Our concern is not with Bohr or Einstein; it is with realism and anti-realism. The issue is not what these people really thought. I have just been using these men and their provocative ideas to get your attention on the question of nature and knowledge. The real issue is whether there is anything in modern physics that favors the sort of anti-realism claimed by Bohr or the realism proposed by Einstein. Is there anything in the experiments or the principles of quantum mechanics or relativity that indicates that there is (or is not) a real world out there, independent of our observations or thoughts, or anything that indicates that we can (or cannot) know how nature really is?

Of course the physics could be neutral on these questions, supporting neither one claim nor the other. These could be purely philosophical questions. They might even be unanswerable questions, immune to resolution by evidence or logic. Then realism or anti-realism would be a matter of speculation and personal taste (and not worth reading a book about). It is also possible that the physics presents mixed evidence about realism, supporting realism in some cases but anti-realism in others. All of this is possible, but none of it is the case. The issue of realism is not a matter of taste, and the physics of quantum mechanics and relativity is neither neutral nor mixed.

The truth about realism is neither entirely as Bohr claims nor entirely as Einstein claims. The truth of the matter is a kind of hybrid of each, and the contents of the physics we have done point to the proper compromise. What we want to do is, under the influence of modern physics, take what is right from the anti-realist ideas and what is right from the realist, and put these together. From the anti-realist, and from the details of quantum mechanics and relativity, we must acknowledge that humans have an unavoidable influence on everything they observe and know. Information does not flow into our minds in a pure, unaltered stream. We change things physically, in the act of observing, and we impose our own conceptual scheme, in the act of knowing. There is no way around this. But from the realist, and again from the details of quantum mechanics and relativity, we must conclude that we can know how nature is, at least in part, independent of our influence. We must be cautious about the optimism of realism. The we-can-know-the-truth attitude comes with a burden of proof that must be carried in each case of claiming to know how nature is. It is a case-by-case responsibility, tied to the specifics of each claim, and requiring a specific kind of realism. We can know some things about how nature is, but perhaps not everything. I like to think of this as a realistic kind of realism, realistic in the sense of being pragmatic and reasonable.

TWO KINDS OF QUESTIONS: METAPHYSICS AND EPISTEMOLOGY

Science is like a microscope, and nature is the specimen. Unlike the optical tool though, science is both a physical and conceptual apparatus. But like a microscope, the task of science is to magnify the specimen and bring it into focus so that its parts and processes can be described and explained. The elements of a microscope are lenses and apertures. The elements of a science are its theories and concepts and terms, as well as its experimental equipment. All of the elements are adjustable, and the business of science is to fine-tune the settings to produce the clearest possible image.

On this analogy of science, epistemology is the inquiry about the microscope itself. When we ask how it works or whether it is presently in focus, we are asking epistemological questions. Questions about the magnification or resolving power of the apparatus are in the same category. And these are all steps to the most pressing epistemological question, How do we know that the image we see in the microscope resembles the specimen itself? Questions about the specimen itself, questions that do not involve our imaging of it, are metaphysical. What properties does this thing have? How does it behave? What happens when this or that force is applied? What laws govern its interaction with other things? These are questions of metaphysics.

The separation between metaphysical and epistemological issues matches the distinction between the two claims made by Bohr. There is an anti-realism about epistemology and an anti-realism about metaphysics. The two concerns are separate but related.

Anti-realism about epistemology casts doubt on the reasons for thinking that the image resembles the specimen. We certainly know what the image looks like. There is no question there. We can describe the image and verify the description. We can tell when the image is clear and sharp. In science, that is, we can tell when things make sense, when the pieces of the description fit together and present a coherent, explanatory account of nature. But, getting back to the microscope analogy, we cannot know if the image resembles the specimen itself, because we cannot see the specimen itself and so we cannot compare one to the other.

We can fiddle with the apparatus, adjusting theories and concepts until the image is crystal clear. But clarity does not mean things are in focus; it just means we fiddled until it made sense to us. This is a psychological achievement, and making-sense-to-us is not the same as being true. Making sense is not even a necessary indication of truth.

Since we cannot possibly know what the specimen looks like without the apparatus, there is no direct way to calibrate and verify the focus of the apparatus. We can know about the image, that is, what the specimen looks like under the influence of the apparatus, how the specimen appears to us, but we cannot know about the specimen as it *is*, independent of this influence. This is epistemological anti-realism, and it is Bohr's claim that physics itself, and quantum mechanics in particular, force this retrenchment in the responsibilities of scientific knowledge.

Metaphysical anti-realism is about the specimen itself. Simply, there is no specimen. There is no quantum world. There is only a hollow container, a reflection of ourselves. There is only the image, and as we change the setting on the microscope, turn this knob or alter that theory, everything changes. As the image changes, so too does the universe change. This kind of anti-realism, by Bohr's reckoning, is also forced upon us by modern physics.

It is worth considering the logical relations between metaphysical and epistemological realisms before separating them and asking if they are supported by the physics. First, a non-relation. Epistemological anti-realism (we cannot know how nature is) does not entail metaphysical anti-realism (there is no independent way that nature is). It is certainly possible to claim that there is an independent reality out there but we cannot know about it. It is also possible to withhold judgment on the metaphysical issue. In fact, this metaphysical agnosticism seems to be a requirement of epistemological anti-realism. This is the relation of note between the two. Epistemological anti-realism precludes metaphysical anti-realism; indeed, it precludes metaphysical anything. If we cannot know how things are (and we cannot, by the standards of epistemological anti-realism) then neither can we know how things are not. If we cannot describe how nature *is*, we cannot describe how nature *is not*. Metaphysical anti-realism, if it is to be believed with any confidence and justification, must presuppose some degree of epistemological realism.

We will sort this out under the influence of the physics itself. It is too important to be left as a conceptual conversation, abstractly about reality, or Reality, or "reality." An informed answer to questions of realism requires that we pull out the manual on the microscope and figure out just how the thing works. With this we need to look closely at the image and note carefully its appearance. And we ought to fiddle with the knobs and even poke at what we think is the specimen. These are the sorts of details that will license a credible resolution of the issues of appearance and reality.

The point is to use the scientific results and scientific standards to draw philosophical conclusions. The physics is to be evidence for a philosophical theory about realisms. We began, in Chapters 1 and 2, with some initial theorizing, that is, some general philosophy. This is how science works. One needs to know what to look for and this requires some selection of important properties and categories of things to work with. This sets the language in which to report and interpret the evidence. Now that the evidence is in and that we have surveyed modern physics, the initial theorizing can be revised, finetuned, and given authority by the findings.

The project here is not to establish the grounds for believing that quantum mechanics and relativity are true. The idea is to use quantum mechanics and relativity as grounds for endorsing a form of realism or anti-realism. The question is, If you believe quantum mechanics and relativity, that is, if you believe modern physics, then what can you say (what *must* you say) about our ability to know how nature is and about the independent existence of things? We are in that sort of conversation that has a line like, "Well, quantum mechanics (or relativity) says that ...," where the "says that" part is something about subjectivity and objectivity, or appearance and reality.

THE METAPHYSICAL ISSUE

Having separated and articulated the two issues of metaphysical and epistemological realism, we can answer them one at a time. First, the metaphysical issue.

There is nothing in the results of modern physics that warrants a claim of metaphysical anti-realism. Any links between quantum mechanics or relativity and metaphysical anti-realism are based on one of two fallac-ies. One fallacy is the inference from facts about the characteristics of *some* things to a claim about the characteristics of *all* things. The other fallacy is the inference from the fact that things do not behave in a familiar or expected way to the claim that these things simply do not exist.

Consider first the fallacy of jumping from some to all. This is what is going on in the obviously misrepresentative summary of the special the-ory of relativity that says that everything is relative. It is also at work in any claim to the effect that the quantum world has no determinate prop-erties and evolves in an entirely indeterministic way. All of these sum-maries of modern physics are false, and all follow the same fallacy. Ac-cording to the special theory of relativity, *some* properties are relative, but some properties are not relative. According to quantum mechanics, *some* properties are indeterminate, but some are not. In both cases it is not just that there is evidence that some things are this way and no evidence that all are. There is in fact proof that not all things are this way. The infer-ence from some to all, in other words, is not a case of stereotyping a whole group on the basis of knowledge of a few and ignorance of the others. It is a case of actually ignoring the manifest counterexamples, that is, ig-noring what you do not want to be true.

Modern physics does warrant the following claim about specific fea-tures of nature, a claim that is clearly relevant to metaphysical anti-realism. Some aspects of nature (and we can specify which ones) are determinate only with respect to a reference frame or an act of measure-ment. This is a long way from claiming that all aspects of nature have this indeterminacy. It is not about all of reality, or Reality, or all of Nature, or the World. The moral of the story is to be specific.

Then there is the other fallacy, of inferring from unfamiliar to nonexis-tent. It too has an analogy to issues of morality, like the stereotyping in the previous fallacy. It is like observing people of another culture and not-ing that they do not participate in *our* morals, and thereby concluding that they have no morals at all. This is the kind of reasoning behind some of the claims that there is no quantum world.

Both quantum mechanics and relativity use familiar concepts. The prop-erties they seek out, the ways they group things together as being the same kind of thing, and the language they employ are all, by and large, classical. The initial theorizing that directs us in what to be looking for attends to such properties of things as their length, position, velocity, and the like. Both quantum mechanics and relativity then find that many of these concepts, these ways of looking at and describing the world, fit only in a rather awkward and complicated way. Some of these properties must be applied as two-place relations. Some of them are indeterminate until measured. Some pairs of them are complementary. The straightforward properties that we are used to (our morals) do not apply in such a straightforward way to the relativistic and quantum worlds. This does not warrant the conclusion that no properties apply. Properties that we thought we understood and that we thought applied to things in simple, absolute, determinate ways, apply only in complicated, unexpected ways. This is no proof at all that there simply are no properties.

THE EPISTEMOLOGICAL ISSUE

The fact that we can say anything at all that is substantive about metaphysical realism or anti-realism is proof of some degree of epistemological realism. The statements about the nature of properties like length, position, or spin orientation are not just groundless speculation. They are not based on anyone's personal predilection. They are based on physics. Physics, in other words, is the source of evidence for these metaphysical claims, these claims about the way that nature is.

But let's not get carried away. The key to a responsible and accurate response to the issue of realism in the metaphysical case was to be specific. This guideline is inherited by epistemological realism. Be specific. Physics can tell us about some things in nature, but not necessarily about everything. The question of realism is of our ability to know more about the world than just the contents of our own immediate observations. Realizing that this need not be an all-or-nothing issue suggests the possibility of an answer that is neither recklessly optimistic nor restrictively pessimistic. Call it **realistic realism**.

We can not know everything, but we can know a lot.

For example, we can know about spin orientation. We can know that it is in fact indeterminate until it is measured. This claim is made under the authority of quantum mechanics, and of course the credibility of quantum mechanics could change. Quantum mechanics could some day be the old, disreputable way of looking at things. But this is another issue. Our concern is what modern physics allows us to say about the world. If quantum mechanics says anything about epistemological realism, that is, about our ability to know how nature is, it says that we can know what things are like even when they are not observed. We can know how nature is, at least in some respects.

Bell's experiment and argument about spin orientation support realistic realism by demonstrating that we can make justified claims about what the world is like even in aspects that we are unable to observe. We can know some things that transcend appearances. Using what we observe in the world as evidence, the challenge of realism is to expand on the evidence to know more. It is a challenge to use the evidence as justification for claims about what is not evident. The accomplishment would be justification, not certainty, and only about some things, not everything. This would be realistic realism.

Bell's work clearly indicates that realistic realism is possible, since it is a proof (justification) about the nature of the particles when no one is looking. These results describe aspects of the objects in a situation in which no one *could* look, that is, before the measurement. If these quantum mechanical results prove anything, they prove that the nature of the particles when they are uninfluenced by us is indeterminate with respect to spin orientation. They are in fact non-partisans, every one. It is not that we cannot tell whether they are spin-up or spin-down. It is not that their spin orientation is uncertain, reflecting an impediment of our knowledge about them. The result of the quantum mechanics is not a counsel of agnosticism about pre-measurement spin orientation, as would suit the anti-realism that limits knowledge to what can be observed. No, Bell's results make definitive claims about unobservables. Now we know how the quantum world (at least this aspect of it) is. The indeterminateness is a feature of the world, not a weakness of our ability to know about the world.

Another approach to the issues of epistemological realism is through the concept of objectivity. Objectivity, if it is to be a useful concept, is best invoked as a feature of knowledge rather than of the things known or studied. It is an aspect of epistemology, not metaphysics. It makes more sense, in other words, to ask whether the activities of quantum mechanics, or astronomy, or perhaps science in general, are more or less objective, more sense than asking whether electrons or the moon or the "quantum world" is objective. The virtue of being objective, when applied via metaphysics and credited to these entities, is merely redundant with the concept of existence. It is in this sense a hollow label to apply after the important work is done, since evaluating this kind of objectivity relies on the relevant existence claims, claims that ought themselves to be justified by an appeal to objectivity of the knowledge. Therefore, claims about objective electrons or an objective moon, presumably constituents of an objective reality, will not contribute to the understanding of the relationship between ourselves as knowers and the world we claim to know.

The more useful, accessible concept of objectivity is purely epistemological. Evaluation of objectivity in this sense will not require comparing a claim about an object in the world to the object itself as it is uninfluenced by the knower. It will not ask, in other words, to compare the view from here against, as Thomas Nagel puts it, "the view from nowhere." We have no access to the view from nowhere, and a useful concept of objectivity must be amenable to evaluation on the basis of wholly accessible information, that is, wholly knower-influenced information. In terms of the microscope, judging its focus must be done entirely on the basis of understanding the features of the image and the microscope itself. We can make no appeal to the features of the specimen.

The accomplishment of objectivity is not in escaping our own point of view (because that is impossible) but in understanding our point of view. The point is not to eliminate the effects of our perspective but to understand how our perspective affects the view, and this can result in knowing which features of the appearance and description are in fact in the world, even though both contributions, from the knower and the known, are in every view of the world.

Both quantum mechanics and relativity are perfectly suited to this kind of objectivity. Part of the accomplishment of each is to point out the human influence on nature's appearance. The special theory of relativity, for example, reveals the physical influence in terms of the relativity of some properties to reference frame. And since all of our measurements are from one reference frame or another, these measurements are relative to our perspective. To paraphrase Nagel, the view from no reference frame is unattainable. But now we know not only *that* properties like length and time duration differ in reference to different frames, we know *how* they differ. It is not that we influence the appearance in an unknown, unaccountable way. The special theory of relativity supplies the details to give us an understanding of our perspective. It tells us specifically how the microscope is working.

Quantum mechanics is similarly forthcoming about the physical influence of observation on the thing observed. Observation of a quantum system somehow causes an abrupt change in things, precipitating a collapse of the state function. Quantum mechanics is short on the details of how this happens, but we know that is does happen. This is where Bell's proof is pivotal. It is proof that, without our interference by observation, the spin orientation of a quantum object is indeterminate. Our physical influence makes it look different; it makes it look determinate with respect to spin orientation. We have this effect on things, and now we know about it. We must keep this in mind, this distortion in the image.

There are aspects of both Bohr's and Einstein's attitudes in this view of objectivity. Bohr would claim that there is a human influence in everything we can know, and so we can only know the world as it is influenced by us. He acknowledges the influence, in other words, and gives into it. Einstein, on the other hand, believes that we can escape the influence and see aspects of the world free and clear of our own effects on it, thus we can know how nature is and not just how it appears to us. The proper hybrid of these two is to admit to the human influence, as Bohr does to begin with, but nonetheless adopt Einstein's conclusion. Because we can know of the human influence, and understand it, we can also know the natural influence. We can know nature's contribution to the appearance. We can know how nature is, at least in some respects.

This achievement of objectivity is more than just intersubjectivity. We are not saying that science has some degree of objectivity because its claims enjoy broad and diverse agreement throughout the profession. Such agreement is too susceptible to the influence of authority, whether it is from teachers, textbooks, or editors, to be of much value in judging whether the agreed claims are likely to be true. The kind of objectivity though that is based on articulating and understanding the human influence on observation and knowledge is an objectivity based on principles, not people or personalities.

The physics of quantum mechanics and relativity are effective for revealing aspects of the physical influence of people. It is also handy for understanding the conceptual influence. Looking at modern physics, we realize that we always have a choice of language in describing nature. There is no denying then that we influence what is said in describing nature. We decide for ourselves what is important to look at and to think about. We decide which features of nature warrant the distinction of being labeled with special terms. Some aspects of things go unmentioned. Others are announced in boldfaced type in the textbooks. We keep track of the mass of an object, for example, and its velocity and its electric charge. Furthermore, we recognize the product of mass and velocity as a property in itself, and call it momentum. The product of mass and charge though is nothing. There is no special category in our descriptive language for this.

The language we choose, the properties and combinations of properties that we choose to distinguish with terms, has an immediate effect on the simplicity and sense of the resulting description of nature. Indeed, the language is chosen for the sense and simplicity it brings about. But it is still our choice. As a colorful example, we could choose to adopt a term like "grellow," that, when translated into our more normal color language, means green for part of the year and yellow for part of the year. With this new color terminology, the description of some aspects of nature becomes somewhat simpler and more straightforward. It is certainly more economical with words. In this new language, for example, the leaves on an aspen tree do not change color. They are always grellow. In the more traditional language, the description of the color of aspen leaves is quite complicated, having to account for the change between the color in summer and the color in fall.

There is a fact of the matter about aspen trees and the color of their leaves. Metaphysical realism is clearly appropriate here. Furthermore, we can know what the fact of the matter is. This is a claim of epistemological realism. It is appropriate even though our descriptions of aspen trees, what we say about them, are not uniquely determined by nature itself. We can say that the color of the leaves changes during the year, if we choose to use the language of green and yellow. We can say that the color of the leaves does not change during the year, if we choose to use the language of grellow. But once we choose the language our conceptual influence is over. Once the language is chosen, nature reports and the description is determined by the facts of life. If we choose the language of green and yellow then we are forced by the facts to say that the color of the leaves changes. At this stage of the descriptive process it is more than just what we can say about nature; it is what we *must* say.

Modern physics clearly shows the two components of our description of nature, ours and the world's. We can choose our geometric conventions to describe spacetime as flat or curved. We can choose what properties of a quantum system to attend to and to measure. But once the choice is made there are things we have to say. The description is constrained by nature. If we choose to talk in terms of spin orientation then we are forced to say that it is indeterminate before the observation.

The human influence on observation and knowledge is unavoidable and irreducible. It has both a physical and a conceptual aspect. We suspected this to begin with, and now the details of modern physics show the human influence and how to deal with it.

IT'S NOT THE END OF THE WORLD

Quantum mechanics has always been a bother to scientific realism. Any theory that elevates uncertainty to a principle would seem to have the limits of knowledge collapse inward, leaving only knowledge of appearance and not reality. Particularly since Bell's proof that local hidden-variable theories are incompatible with the results of quantum mechanics, the quantum theory has been seen as an obstacle to any scientific realism that would allow knowledge to transcend the observable. What with spooky actions at a distance and entities that are somehow waves and particles yet neither, the concept of quantum reality seems almost an oxymoron. But we have seen that this is wrong. The pessimism about knowledge is unfounded, and it is exactly a proof like Bell's that gives reason for optimism. It is a proof about how quantum things are when we are not, and could not be, looking. It is proof that things are not as they appear.

Relativity might also seem to be in the way of realism. There is the temptation to slide from relativity to relativism where there is no unique truth or where everything we can know about the truth is relative to ourselves as individuals. But this inference is not only unwarranted, it is opposed by relativity. Some features of nature are absolute and others are relative to a reference frame. These are the facts. And once the reference frame is specified, the relative properties have determinate values. It has nothing to do with people.

In general, modern physics describes the world in terms that challenge our imagination and intuition. But that's our problem. It's not the end of the world. It's not even the end of our knowledge of the world. It's just the end of the world as we knew it.

REFERENCES AND SUGGESTED READING

PHILOSOPHICAL BACKGROUND

These books are all relevant to the issues of appearance and reality, and the human influence on observation and knowledge. They are all general accounts of philosophy of science or of epistemology and metaphysics.

Giere, R. *Explaining Science*. Chicago: University of Chicago Press, 1988. This is an accessible work on philosophy of science that adheres to both naturalized epistemology (that is, using the facts of science to guide the description of how we know about nature) and realism.

Goodman, N. *Ways of Worldmaking*. Indianapolis: Hackett, 1978. This is a provocative and charming book that stresses the human contribution to the description of nature. It is the inspiration for my closing analogy about grellow trees.

Hesse, M. *The Structure of Scientific Inference*. Berkeley: University of California Press, 1974. The first chapter, "Theory and Observation," is a clear description of the reciprocal relation between observation and theory in science.

Kant, I. *The Critique of Pure Reason*. Translated by N. Kemp Smith. New York: Macmillan, 1929. Philosophers will recognize the Kantian influence in much of the epistemology in my account of modern physics, particularly the approach to observation and the human contribution. Kant is difficult to read, but his ideas are a landmark of modern philosophy.

Kosso, P. *Reading the Book of Nature*. Cambridge: Cambridge University Press, 1992. This is a short and easy book on the philosophy of science.

Leplin, J., ed. *Scientific Realism*. Berkeley, University of California Press, 1984. This is an anthology of philosophers with various arguments for and against realism.

Miller, R. *Fact and Method*. Princeton, NJ: Princeton University Press, 1987. Long and not always easy, this is a thorough and novel argument for realism in science. It closes with a chapter on quantum mechanics and realism.

Nagel, T. *The View from Nowhere*. Oxford: Oxford University Press, 1986. A book about the human perspective, this is not just about epistemology but about ethics as well.

Quine, W.V.O., and J. Ullian. *The Web of Belief*, 2d ed. New York: Random House, 1978. This is a very readable book about the network structure of knowledge.

Scheffler, I. *Science and Subjectivity*. Indianapolis: Bobbs-Merrill, 1967. Contrary to its title, this is more about objectivity in science. It pays particular attention to our subjective contribution in choosing the descriptive language and to the world's determination of what must be said in that language.

RELATIVITY

These are the books that helped me to understand and explain the special and general theories of relativity and the more philosophical details of space and time. They are all clearly written and rich in more details about relativity.

Ellis, G., and R. Williams. *Flat and Curved Space-times*. Oxford: Clarendon Press, 1988. There are lots of good textbooks on relativity, but this one is a winning combination of clarity, depth, and intuitive approach. It is a particularly good source of information on the general theory of relativity.

Mook, D., and T. Varish. *Inside Relativity*. Princeton, NJ: Princeton University Press, 1987. This one is good for the special theory of relativity. The explanation is clear and well illustrated.

Reichenbach, H. *The Philosophy of Space and Time*. Translated by Maria Reichenbach and John Freund. New York: Dover, 1958. Measurement of space and time and the conventionality of geometry are treated with insight and clarity in this classic.

Sklar, L. *Space, Time, and Spacetime*. Berkeley: University of California Press, 1974. This is a good source of information on the history and philosophy of such issues as substantival versus relational space, the direction of time, and the conventionality of geometry and topology.

Taylor, E., and J. Wheeler. *Spacetime Physics*. San Francisco: W.H. Freeman and Company, 1963. This textbook is not always easy to read but it develops a deeply intuitive account of relativity. It has a large and helpful supply of worked examples. It is better for special relativity than general.

QUANTUM MECHANICS

This is a tiny sample from the many works on quantum mechanics and its philosophical ramifications. These are chosen for their clarity and their careful approach to interpreting the theory.

Albert, D. *Quantum Mechanics and Experience*. Cambridge, MA: Harvard University Press, 1992. This is a cheerful and clear introduction to the facts and interpretations of the EPR experiment and Bell's proof. It is also a good source for Bohm's theory.

Bohm, D. *Quantum Theory*. Englewood Cliffs, NJ: Prentice-Hall, 1951. This is a classic textbook by one of the early contributors to the development of quantum mechanics. Bohm is responsible for the setup of the EPR experiment we used.

Herbert, N. *Quantum Reality*. New York: Doubleday, 1987. Written for the reader with no background in mathematics or physics, this explores a variety of metaphysical consequences of the quantum theory.

Mermin, N.D. *Boojums All the Way Through*. Cambridge: Cambridge University Press, 1990. No one can explain the difficult details of quantum mechanics and Bell's proof in particular, with such clarity, insight and charm as David Mermin. My description of Bell's proof owes a lot to Mermin. This book is a collection of his essays on the quantum theory and its interpretation.

Rae, A. *Quantum Mechanics*, 2d ed. Bristol, England: Adam Hilger, 1986. This is one of the best introductory textbooks, written by a physicist with an interest in the philosophical issues.

Rae, A. *Quantum physics: illusion or reality?* Cambridge: Cambridge University Press, 1986. This is the best survey on quantum mechanics and its philosophical consequences and questions written for the general reader. It is short and sweet.

HISTORICAL AND BIOGRAPHICAL ACCOUNTS

My account of modern physics has ignored the historical development of the ideas and the people who developed them. It would be wise to supplement what you have read here with some of the following books that put the ideas back into their historical setting.

Gamow, G. *Thirty Years that Shook Physics*. Garden City, NY: Anchor Books, 1966. Always amusing, this story of the people and events in the beginning of the quantum theory is worth reading just for the fun of it. Here is where find out about Bohr's life in the Carlsberg brewery, the home of the Copenhagen interpretation.

Honner, J. *The Description of Nature*. Oxford: Oxford University Press, 1987. This might just as well fit in the section on philosophical background. It is about Bohr and Einstein and the points of agreement and disagreement in their philosophies of physics.

Pagels, H. *The Cosmic Code*. New York: Bantam Books, 1982. This could just as well be in the section on quantum mechanics because it is an insightful and sensible survey of the key ideas. The development of the topic is historical, and so it is a good place to find the ideas in context.

Pais, A. *Subtle is the Lord*. New York: Oxford University Press, 1982. This is a wonderfully written portrait of Einstein by an author who understands and can explain the big ideas.

Sachs, M. *Einstein versus Bohr*. La Salle, IL: Open Court, 1988. As we have done, Sachs uses Einstein and Bohr to focus on the issue of realism in modern physics. His account though is much more forthcoming with details on the historical context and the motivations and ideas of the men themselves.

PRIMARY SOURCES

You might want to decide for yourself what Bohr or Einstein thought about physics or realism. If that is the case then you should read what they actually said. Here are some suggested sources of writings by the physicists whose ideas we have been talking about.

Bell, J. *Speakable and Unspeakable in Quantum Mechanics*. Cambridge: Cambridge University Press, 1987. This is a collection of Bell's articles on philosophical aspects of quantum mechanics. It has his original presentation of Bell's proof.

Bohr, N. *Atomic Physics and Human Knowledge*. New York: Wiley, 1963. This is a collection of essays.

Einstein, A. *Relativity: The Special and the General Theory*. Amherst: Prometheus Books, 1995. This is Einstein's presentation of the development and consequences of relativity, written without mathematics. It is for a general audience.

Einstein, A., et. al. *The Principle of Relativity*. New York: Dover, 1952. This is a collection of original papers by Einstein, Minkowski, and others on the special and general theories of relativity. It has Einstein's example of the rotating reference frames that we used to introduce the idea of the curvature of spacetime.

Mach, E. *The Science of Mechanics*. La Salle, IL: Open Court, 1960. This is where to find the original presentation of Mach's principle on the relational model of space.

INDEX

●